T0181917

Synthese Library

Studies in Epistemology, Logic, Methodology, and Philosophy of Science

Volume 381

More information about this series at http://www.springer.com/series/6607

Alexander Gebharter

Causal Nets, Interventionism, and Mechanisms

Philosophical Foundations and Applications

 Springer

Alexander Gebharter
Department of Philosophy
University of Düsseldorf
Düsseldorf, Germany

D61

Synthese Library
ISBN 978-3-319-84270-7   ISBN 978-3-319-49908-6   (eBook)
DOI 10.1007/978-3-319-49908-6

Printed on acid-free paper

This Springer imprint is published by Springer Nature
The registered company is Springer International Publishing AG
The registered company address is: Gewerbestrasse 11, 6330 Cham, Switzerland

Contents

Chapter 1
Introduction

Abstract This chapter provides a brief introduction to the book as well as an overview of the subsequent chapters. The book consists of two main parts: A more theoretical part on the philosophical foundations of the theory of causal Bayes nets and a more application oriented part. The former part (Chap. 4) supports the theory of causal Bayes nets by an inference to the best explanation. It can also be seen as a response to David Hume's skeptical challenge about whether causation is something ontologically real. The latter part is more application oriented. In particular, I try to shed some new light on Woodward's (2003) interventionist theory of causation (Chap. 5) and the new mechanist debate (Chap. 6). Chapters 2 and 3 are of introductory nature. Chapter 2 presents the formal preliminaries needed for subsequent chapters; Chap. 3 provides a self-contained introduction to the causal Bayes net framework.

Before some introductory words and a brief overview of this book, let me say a few words about who might and who might not enjoy reading this book. This book is mainly written for empirically oriented philosophers interested in theories of causation, the metaphysics of causation, or in what one can learn from applying an empirically informed theory of causation to related topics in philosophy (such as mechanisms). It is intended to contribute to several philosophical issues and discussions in novel ways. It uses and heavily relies on basics of the causal Bayes net framework. The book, however, is not intended to further develop this framework as a methodological tool for causal discovery, inference, etc. It might, hence, not be very interesting to researchers mainly interested in the further development of causal Bayes nets.

In the book I argue that the causal Bayes net framework—which is often only thought of as a nice to have methodological tool for causal inference—actually can be axiomatized in such a way that one gets a full (though metaphysically light-wight) theory of causation. I then argue that this particular theory of causation has clear advantages over its more classical rivals on the market from an empirical point of view: The theory can be backed up by an inference to the best explanation of certain phenomena and several versions of the theory (i.e., combinations of axioms) lead to empirical consequences by whose means the theory as a whole can be tested. In these respects, the theory behaves exactly like a successful empirical theory of

© Springer International Publishing AG 2017

A. Gebharter, *Causal Nets, Interventionism, and Mechanisms*,
Synthese Library 381, DOI 10.1007/978-3-319-49908-6_1

the sciences, and this is—as far as I know—something novel for a philosophical theory. I also argue that the theory can meet Hume's critical challenge: Actually, we have good reasons to believe that causation is something ontologically real out there in the world.

Another point I investigate in this book is how Woodward's (2003) prominent interventionist theory of causation, which is clearly inspired by Pearl's (2000) work on causal Bayes nets, actually relates to the theory of causal Bayes nets. I hope that philosophers interested in causation and in theories of causation (such as Woodward's interventionist theory) will find these topics interesting.

The book might also be of interest to philosophers working on mechanisms. It collects most of my former work on the representation of mechanisms within a causal Bayes net framework and presents this work and its possible philosophical consequences for the mechanism debate within one single chapter. It also contains my responses to Casini's (2016) recent defense of the modeling approach developed by Casini et al. (2011) and objections to my approach. It also investigates in how far my modeling approach allows for a representation of constitutive relevance relationships. The book also sketches a novel suggestion of how to discover constitutive relevance relationships that does not rely on interventions or strict regularities. Summarizing, what I do in the book can be seen as an appeal to philosophers: The theory of causal Bayes nets is a real alternative to more classical theories of causation, especially when one is interested in solving philosophical problems related to causation in an empirically informed way.

Finally, this book also features a self-contained and easily accessible short introduction to causal Bayes nets. This might be helpful since many philosophers interested in causation and related topics such as mechanisms are still not familiar with that framework. Readers already familiar with the framework can simply skip Chap. 2 and also large parts of Chap. 3.

Let me now start the general introcution with some historical remarks. In particular, I would like to start with Hume's thoughts about causation. Since Hume's (1738/1975; 1748/1999) striking critique on causation as something ontologically real, causation was more or less stepmotherly treated by analytic philosophers for a long time. In particular, Hume argued that we cannot have any knowledge about causation as a relation out there in the world. Cause-effect relations between events are neither observable, nor do we have good reasons to infer their existence. All that we can observe are sequences of events: u occurs before v occurs. If events like u are regularly followed by events like v and if we become aware of this regularity, then we begin to anticipate events of the same kind as v when we observe events of the same kind as u. We may refer to u-events as causes of v-events in that case, but there is no evidence for the existence of something ontologically real that corresponds to the postulated causal connections between u-events and v-events. All that we can know to be out there in the world are regularities. So, the Humean might argue, it is quite likely that postulating causal relations in addition to regularities is nothing more than an evolutionary successful subjective feature of our minds that allows for a cognitively sparse structuring of our environment.

Hume's (1738/1975; 1748/1999) skeptical stance on causation was adopted and taken to extremes by positivistic philosophy of science and empirically oriented analytic philosophers. According to Carnap (1928/2003, pp. 263ff), for example, causation is a purely metaphysical concept and any causal talk should be eliminated from philosophy and the sciences. Russell (1912, p. 193) even went a step further when he famously claimed that "the law of causation [...] is a relic of a bygone age, surviving, like the monarchy, only because it is erroneously supposed to do no harm". From the 1970s on, however, causation has become more and more discussed in the literature again. For evidence, see Williamson (2009, sec. 8), who provides proportions (taken from the British Library database) of books published in English language whose titles include words beginning with 'caus-'. One plausible explanation for this trend Williamson considers is the influence of probabilistic theories of causation (e.g., Cartwright 1979; Reichenbach 1956/1991; Suppes 1970) and especially the influence of the development of the causal Bayes net formalism (Neapolitan 1990, 2003; Pearl 1988, 2000; Spirtes et al. 2000).

The latter does not explicitly define causation, like it is typically done in more traditional philosophical theories of causation. It is rather oriented on the success of empirical theories: It connects causal structures to empirical data (in the form of probability distributions) by several conditions. That this is a step in the right direction can be seen by the framework's usufulness; it can be successfully applied to many scientific domains. The framework provides a multitude of powerful inference methods for uncovering causal structures on the basis of empirical data for different situations and presupposing different causal background knowledge (see, e.g., Spirtes et al. 2000). It also allows for generating models which can be used for providing causal explanations and predictions, for testing causal hypotheses, and even for computing the effects of possible interventions on the basis of non-experimental data.

From an empirical and application oriented point of view, the causal Bayes net framework seems to be a highly promising starting point for developing a useful and general theory of causation. However, Hume's (1738/1975; 1748/1999) skeptical challenge is still uncountered. Hume may still be right that postulating causal relations in addition to regularities may be nothing more than a subjective feature of our minds. One part of this book (*viz.* Chap. 4) can be seen as an encounter with Hume's challenge. It is based on Schurz and Gebharter (2016). In Chap. 4 I argue that there are, contrary to what Hume claimed, good reasons to assume that there is something objective and mind-independent that corresponds to causal relations (provided causal relations are characterized as they are within the theory of causal nets). The basic underlying idea is that metaphysical concepts behave analogously to theoretical concepts in scientific theories. They are not defined, but only implicitly characterized by the theory's axioms.

Metaphysical concepts have, just like (good) theoretical concepts, to satisfy certain standards (cf. Schurz 2013, ch. 5): They should unify empirical phenomena of different domains, allow for building models/hypotheses that generate empirically testable predictions, provide the best explanation of certain empirical phenomena, and they should provide empirically testable predictions for the whole theory, i.e., empirical consequences by whose means not only the theory's models, but also the theory as a whole becomes empirically testable. The more of these criteria a theoretical concept satisfies, the more plausible it becomes that the postulated theoretical concept corresponds to something out there in the world.

Now the idea that metaphysical concepts behave like theoretical concepts can be seen as a new approach to metaphysics, as a modern version of an inductive metaphysics which makes heavy use of inference to the best explanation. Metaphysical concepts are introduced only when required for explaining otherwise unexplainable empirical phenomena. They are axiomatically characterized and they are only characterized as far as it is demanded by the empirical phenomena these metaphysical concepts should explain. Yet this does not suffice for a good metaphysical theory. It must also be investigated whether the theory resulting from an inference to the best explanation as a whole does have empirical content. Only if also the answer to this latter question is a positive one can the theory prove its worth.

While Hume (1738/1975, 1748/1999) saw no reason to assume that causation is something independent of our minds, we nowadays—thanks to modern post-positivistic philosophy of science—possess methods for evaluating the plausibility of the existence of postulated theoretical entities. And if (good) metaphysical concepts behave like (good) theoretical concepts, these methods can also be applied to metaphysical concepts. So if our venture succeeds, we can reply to the Humean: "If you belief in the existence of the theoretical entities postulated by successful empirical theories (such as *force* in Newtonian physics), then you better also belief in causation as something ontologically real."

So one part of this book, *viz.* Chap. 4, is of fundamental and theoretical nature. It provides several reasons for why a theory of causation based on the causal Bayes net framework gives us the best grasp on causation we have so far from an empirical point of view. The following two parts Chaps. 5 and 6 are more application oriented. They make use of some of the results obtained in Chap. 4 and demonstrate how the theory of causal nets can be endorsed to shed new light on more philosophical topics from an empirically informed point of view.

In Chap. 5 I develop a novel reconstruction of Woodward's (2003) interventionist theory of causation—one of the most prominent theories of causation at the moment—within the theory of causal nets. The endeavor of reconstructing Woodward's theory within the theory of causal nets allows one to see in which respects the two theories agree and in which respects they diverge from each other. It also allows for uncovering several weak points of Woodward's theory which may otherwise have been overlooked. I highlight some of these weak points of Woodward's interventionist theory of causation and suggest several modifications of the theory to avoid them. This will result in two alternative versions of an interventionist theory. The basic causal notions of these alternative theory versions will turn out to provably coincide with the corresponding causal notions within the theory of causal nets under suitable conditions.

In the second more application oriented part (i.e., Chap. 6) I enter the new mechanist debate (see, e.g., Craver 2007b; Glennan 1996; Machamer et al. 2000) within the philosophy of science. Mechanists claim that all (or almost all) explanations should consist in pointing at the phenomenon of interest's underlying mechanism. Mechanisms are systems which connect certain inputs to certain outputs. To mechanistically explain a phenomenon means to uncover the causal micro structure of the system that mediates between such inputs and outputs. In Chap. 6 I am mainly interested in the question of how mechanisms can be represented within the theory of causal nets. I discuss a proposal how to model mechanisms made by Casini et al. (2011). I present three problems with Casini et al.'s approach first presented in Gebharter (2014, 2016). I then present my alternative approach and compare it with Casini et al.'s approach. I discuss Casini's (2016) recent defense against two objections to Casini et al.'s approach made in Gebharter (2014) and also his recent objections to my account. I then make a suggestion how constitutive relevance relations (cf. Craver 2007a,b) can be represented within my modeling approach for mechanisms. The rest of Chap. 6 is based on Gebharter and Schurz (2016): Clarke et al. (2014) further develop Casini et al.'s modeling approach for mechanisms in such a way that it can also be applied to mechanisms featuring causal feedback. I show that also my alternative modeling framework can be modified in a similar way. I compare the two frameworks and it will turn out that one advantage of my alternative approach over Clarke et al.'s approach is that causal models do not only allow for observation based predictions, but also for computing post intervention distributions on the basis of non-experimental data.

Structuring This book is structured as follows: The three parts already described above (i.e., Chaps. 4, 5, and 6) are preceded by two parts which are to large extents of introductory nature. In Chap. 2 I introduce the formal preliminaries required for the causal Bayes net framework and the remaining chapters. It contains a brief introcution to probability theory (Sect. 2.3), of variables and probability distributions over sets of variables (Sect. 2.4), probabilistic dependence and independence relations (Sect. 2.5), graphs (Sect. 2.6), and Bayesian networks (Sect. 2.7). Everyone already familiar with these concepts can simply skip this part. The next chapter (i.e., Chap. 3) presents the basics of the causal Bayes net formalism. It introduces causal graphs and causal models (Sect. 3.2) and the framework's core conditions: the causal Markov condition (Sect. 3.3), the causal minimality condition (Sect. 3.4.1), and the causal faithfulness condition (Sect. 3.4.2). But there are also some novel contributions to be found in Chap. 3. For each of the conditions mentioned I present an alternative and philosophically more transparent version. I also briefly discuss a weaker version of the faithfulness condition. I also suggest and justify axioms corresponding to the minimality condition and the mentioned weaker version of the faithfulness condition. Together with the axiom corresponding to the causal Markov condition to be developed in Chap. 4 these axioms can be seen as the core of a philosophical theory of causation based on the causal Bayes net formalism.

The next three parts—already mentioned above—are the main parts of this book. They are structured as follows: Chap. 4 starts with the theory of causal nets' explanatory warrant (Sect. 4.2) and discusses several empirical phenomena which require an explanation. Driven by an inference to the best explanation of these phenomena, step by step an axiom is developed that postulates binary causal relations which are characterized by the causal Markov condition. In Schurz and Gebharter (2016) we only gave a sketch of how such an inference to the best explanation may proceed. In that paper we investigated only simple causal systems with three variables and without latent common causes. In Sect. 4.2 I also consider more complex causal systems in the possible presence of latent common causes. The next big part of Chap. 4 is about the theory of causal nets' empirical content. I present and briefly discuss several theorems from Schurz and Gebharter (2016) and also add new theorems.

I start Chap. 5 by presenting Woodward's (2003) interventionist theory of causation (Sect. 5.2.1) and focus on the theory's four central notions: total causation, direct causation, contributing causation, and intervention. I also consider under which specific assumptions these notions of causation should be expected to coincide with the corresponding causal notions within the theory of causal nets. Then I develop a reconstruction of Woodward's interventionist theory (Sect. 5.2.2). For this reconstruction I have to introduce the concept of a dependence model and the concept of an intervention expansion. This part partially builds on Gebharter and Schurz (2014). In that paper we suggested a reconstruction of Woodward's notion of direct causation within the theory of causal nets. This reconstruction, however, used probability distributions, while the reconstruction in Chap. 5 uses dependence models instead. The latter allow for a reconstruction that is much closer to Woodward's original ideas (for details, see Sect. 5.2.2). I also develop reconstructions of Woodward's notions of total and contributing causation.

After reconstructing Woodward's (2003) three notions of causal relations, I highlight three problems with Woodward's theory: problems which may arise in systems featuring variables one cannot intervene on (Sect. 5.3.1), problems with systems featuring intermediate causes that turn out to be intervention variables (Sect. 5.3.2), and problems with systems featuring deterministic causal chains or deterministic common cause structures (Sect. 5.3.3). I discuss several modifications one could make to Woodward's theory to avoid the three problems, and present two alternative interventionist theories of causation in Sect. 5.4, which are basically the result of accepting the most reasonable of the modifications considered. One of these alternative versions requires interventions to be deterministic (Sect. 5.4.1), the other one also allows for indeterministic (or stochastic) interventions (Sect. 5.4.2). I show that the notions of total, direct, and contributing causation within these alternative versions coincide with the corresponding graph theoretical notions with which they are expected to coincide under suitable assumptions.

The last part of this book, i.e., Chap. 6, starts with a few words on mechanisms (Sect. 6.1). Next I introduce the recursive Bayes net formalism (Sect. 6.2.1) and Casini et al.'s (2011) suggestion of how to use this framework to model mechanisms (Sect. 6.2.2). In Sect. 6.3 I highlight three problems with Casini et al.'s approach: recursive Bayes nets do not provide any information about how a mechanism's input and output are causally connected to its causal micro structure (Sect. 6.3.1), bottom-up experiments cannot be represented by means of intervention variables in recursive Bayes nets (Sect. 6.3.2), and the method for representing and computing post intervention distributions Casini et al. suggest (and Clarke et al. 2014 endorse) does regularly lead to false consequences (Sect. 6.3.3). The first two problems are reproduced from Gebharter (2014), while the third problem, which is also discussed in more detail here, was first introduced in Gebharter (2016). In Sect. 6.4.1 I present an alternative approach for modeling mechanisms developed in Gebharter (2014): the multilevel model approach. I compare this alternative approach with Casini et al.'s approach in Sect. 6.4.2, show that it does not fall victim to the three problems mentioned before, and discuss some possible philosophical consequences in Sect. 6.4.4. I also discuss Casini's (2016) recent objections to my approach and his recent defense of the recursive Bayes net approach. In Sect. 6.5 I investigate the question how constitutive relevance relations can be implemented in models of mechanisms within my modeling approach. I also sketch a novel suggestion of how constitutive relevance relations could be inferred and distinguished from causal relations on the basis of purely observational data.

The last section of this part (Sect. 6.6) is based on Gebharter and Schurz (2016). In this section I expand my alternative approach for representing mechanisms in such a way that it can account for the diachronic mechanisms and that it can also be fruitfully applied to mechanisms featuring causal cycles. I introduce a simple exemplary toy mechanism in Sect. 6.6.1. In Sects. 6.6.2 and 6.6.3 I show how causal cycles can be handled within my alternative modeling approach, and in Sect. 6.6.4 I compare my approach with a similar extension of Casini et al.'s approach developed by Clarke et al. (2014).

Chapter 2
Formal Preliminaries

Abstract This chapter introduces the formal concepts required for subsequent chapters. Some notational conventions are maintained and important terms such as 'relation', 'function', and 'structure' are explicated. The probabilistic concepts relevant for later chapters are illustrated. I also explain the most important graph-theoretical concepts and introduce Bayesian networks. Most of the presented concepts are illustrated by means of simple examples.

2.1 Overview

This chapter introduces the formal concepts required for subsequent chapters. In Sect. 2.2, some notational conventions are maintained and important terms such as 'relation', 'function', and 'structure' are explicated within a set theoretical framework. In Sect. 2.3, the relevant probabilistic concepts as well as some theorems which will be useful in later chapters are illustrated. Statistical variables are introduced in Sect. 2.4; the main differences between statistical variables and predicate constants are highlighted. Next, the notion of a probability distribution over a set of statistical variables is explicated, followed by demonstrating how the in Sect. 2.3 introduced probabilistic concepts and theorems can be applied to statistical variables. In Sect. 2.5 probabilistic dependence/independence relations between statistical variables and some interrelated notions are introduced. In Sect. 2.6 the most important graph-theoretical concepts are explained. Last but not least, Sect. 2.7 introduces Bayesian networks, and thus, finally connects graphs to probability distributions over sets of statistical variables. All theorems and equations presented in Chap. 2 are stated without proof; most of the presented concepts are illustrated by means of simple examples. Readers already familiar with these concepts can just skip the whole chapter.

© Springer International Publishing AG 2017
A. Gebharter, *Causal Nets, Interventionism, and Mechanisms*,
Synthese Library 381, DOI 10.1007/978-3-319-49908-6_2

2.2 Logic and Set Theory

During the following chapters I presuppose a standard first order logic with identity in which the symbols '¬', '∧', and '∨' stand for the standard sentential connectives *negation*, *conjunction*, and *disjunction*, respectively, while '=' stands for *identity*. The symbols '∀' and '∃' stand for *universal* and *existential quantification*, respectively. Because plain arrows are reserved for representing causal relations, the symbols '⇒' and '≡' shall be used for the sentential connectives *implication* and *equivalence*, respectively. Upper-case letters from A to C ('A', 'B', 'C', 'A_1', 'B_1', 'C_1', etc.) are meta-variables for formulae, lower-case letters from a to c ('a', 'b', 'c', 'a_1', 'b_1', 'c_1', etc.) are individual constants, and lower-case letters from u to w ('u', 'v', 'w', 'u_1', 'v_1', 'w_1', etc.) are individual variables. 'Q', 'R', 'Q_1', 'R_1', etc. are used for predicate constants. '±' is used as a meta-symbol so that '$\pm A$' stands for 'either A or $\neg A$', while '$\equiv_{df}$' and '$=_{df}$' are used as meta-symbols that indicate definitions.[1]

In addition I presuppose a typical set theoretical framework in which '∈' stands for the *element relation*, while '{ }' indicates specific sets. Most of the time, 'M', 'N', 'M_1', 'N_1', etc. will be used for designating sets. '∅' stands for the *empty set*. The symbols '⊆' and '⊂' stand for the relations *subset* and *proper subset*, respectively, while '∪', '∩', '⋃', '⋂', '$\mathcal{P}$', '×', '$^{-}$', and '\' are constants for the functions *union*, *intersection*, *general union*, *general intersection*, *powerset*, *Cartesian product*, *complement*, and *relative complement*, respectiveley. '⟨ ⟩' indicates *n*-tuples, '[ ]' stands for *closed* and '] [' for *open interval*, and '| |' for *cardinality*.

2.3 Probability Theory

Basics When specifying a probability function, one typically starts by identifying a set of *elementary events* $e_1, \ldots, e_n$. $e_1, \ldots, e_n$ can, for example, be the possible outcomes of an experiment. Next one can choose an algebra over $\{e_1, \ldots, e_n\}$ such as, for instance, $\mathcal{P}(\{e_1, \ldots, e_n\})$.[2] Let us call $A \in \mathcal{P}(\{e_1, \ldots, e_n\})$ an *event*. Then we can define a *probability function* P as a function satisfying the following three axioms of probability calculus:

(1) $0 \le P(A) \le 1$

(2) $P(A \vee \neg A) = 1$

(3) $P(A \vee B) = P(A) + P(B)$, *provided A and B are mutually exclusive* (2.1)

---

[1] The symbol '$\equiv_{df}$' stands for a definition via equivalence, while '$=_{df}$' stands for a definition via identity.

[2] An algebra over a set M is a subset of $\mathcal{P}(M)$ that contains M and is closed under the complement as well as the union operation.

Fig. 2.1
$$P(A) = P(A \wedge B) + P(A \wedge \neg B)$$

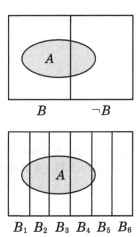

$B \qquad \neg B$

Fig. 2.2
$$P(A) = \sum_{i=1}^{6} P(A \wedge B_i)$$

$B_1 \; B_2 \; B_3 \; B_4 \; B_5 \; B_6$

According to axiom (1), the probability of an event A lies within the interval $[0, 1]$, axiom (2) assures that the probability of the *sure event* equals 1, and axiom (3) tells us how to compute the probability of an event $A \vee B$ on the basis of the probabilities of A and B, provided A and B are mutually exclusive.

Important notions and theorems Let us take a brief look at some interesting stipulations and theorems of probability calculus. Let us begin with the following formula that tells us how to compute the probability of A whenever we know the probabilities of $A \wedge B$ and $A \wedge \neg B$:

$$P(A) = P(A \wedge B) + P(A \wedge \neg B) \tag{2.2}$$

How Equation 2.2 works can be illustrated by means of the diagram in Fig. 2.1. The areas in the diagram correspond to the probabilities assigned by P. The area of the whole diagram corresponds to the sure event which gets probability 1. The left half (area B) of the diagram corresponds to the probability of B (which equals 0.5) and the right half (area $\neg B$) of the diagram corresponds to the probability of $\neg B$ (which also equals 0.5). The part of area A that lies in area B corresponds to the probability of $A \wedge B$, the part of area A that lies in area $\neg B$ corresponds to the probability of $A \wedge \neg B$, and the whole area A corresponds to the probability of A, i.e., the probability of $A \wedge B$ plus the probability of $A \wedge \neg B$.

The basic idea behind determining the probability of A by means of the probabilities of $A \wedge B$ and $A \wedge \neg B$ can be generalized to the so-called *law of total probability* (see Fig. 2.2 for an illustration by means of a diagram). Whenever $\{B_1, \ldots, B_n\}$ is a set of exhaustive and mutually exclusive events, then:

$$P(A) = \sum_{i=1}^{n} P(A \wedge B_i) \tag{2.3}$$

Fig. 2.3 $P(A|B) = \frac{P(A \wedge B)}{P(B)}$

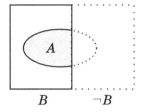

$$B \qquad \neg B$$

An important probabilistic concept is the concept of *conditional probability*, which can be defined for all cases in which $P(B) > 0$ holds[3]:

$$P(A|B) =_{df} \frac{P(A \wedge B)}{P(B)} \tag{2.4}$$

The main idea behind this definition is that the probability of A conditional on B should equal the probability of A in the light of B, i.e., the probability of A when B is treated as if it were the sure event. This can be illustrated by means of the diagram in Fig. 2.3: Like in Fig. 2.1, the area of the whole diagram (i.e., the area included within the continuous and the dashed lines) corresponds to the sure event, which gets probability 1, while the areas B and $\neg B$ correspond to the probabilities of B and $\neg B$, respectively. When conditionalizing on B, we treat B as if it were the sure event, i.e., as if B would get probability 1. We imagine the area of the whole diagram to be restricted to area B. The probability of A conditional on B then corresponds to the ratio of the parts of area A in B to the whole area B.

The so-called *product rule*, a theorem that allows one to compute the joint probability of two events A and B on the basis of the conditional probability of A given B and the probability of B, is a direct consequence of the definition of conditional probability (Equation 2.4):

$$P(A \wedge B) = P(A|B) \cdot P(B) \tag{2.5}$$

The product rule can be generalized to the so-called *chain rule formula*:

$$P(A_1 \wedge \ldots \wedge A_n) = P(A_1) \cdot P(A_2|A_1) \cdot \ldots \cdot P(A_n|A_1 \wedge \ldots \wedge A_{n-1}) \tag{2.6}$$

Or equivalently:

$$P(A_1 \wedge \ldots \wedge A_n) = \prod_{i=1}^{n} P(A_i|A_1 \wedge \ldots \wedge A_{i-1}) \tag{2.7}$$

---

[3]This restriction is required because division by 0 is undefined.

As we have seen, the product rule (Equation 2.5) allows us to compute the joint probability of $A \wedge B$ on the basis of the conditional probability of A given B and the probability of B. But what if we want to know the probability of A? Well, in that case we just have to sum up the conditional probability of A in the light of B weighted on B's probability and the conditional probability of A in the light of $\neg B$ weighted on $\neg B$'s probability:

$$P(A) = P(A|B) \cdot P(B) + P(A|\neg B) \cdot P(\neg B) \tag{2.8}$$

Equation 2.8 can be illustrated by Fig. 2.1: Since, according to Equation 2.5, $P(A|B) \cdot P(B)$ equals $P(A \wedge B)$ and $P(A|\neg B) \cdot P(\neg B)$ equals $P(A \wedge \neg B)$, the part of area A in B in Fig. 2.1 corresponds to $P(A|B) \cdot P(B)$ and the part of area A in $\neg B$ in Fig. 2.1 corresponds to $P(A|\neg B) \cdot P(\neg B)$. The sum of these two parts of A corresponds to the probability of A.

Equation 2.8 can be generalized to the following *law of total probability*. Whenever $\{B_1, \ldots, B_n\}$ is a set of exhaustive and mutually exclusive events, then:

$$P(A) = \sum_{i=1}^{n} P(A|B_i) \cdot P(B_i) \tag{2.9}$$

An illustration of how Equation 2.9 works (analogously to the one given above for Equation 2.8 by means of Fig. 2.1) can be given for Equation 2.9 by means of Fig. 2.2.

The following equation is called *Bayes' theorem*:

$$P(B|A) = \frac{P(A|B) \cdot P(B)}{P(A)} \tag{2.10}$$

Bayes' theorem is a direct consequence of the definition of conditional probability (Equation 2.4) and the product rule (Equation 2.5).

2.4 Statistical Variables

Basics Statistical variables represent properties at a very abstract level and can be used like predicate constants in first order logic. In more detail, a statistical variable is a function that assigns an element of a specified space of exhaustive and mutually exclusive sets of properties to every individual in a given domain D. The sets of properties a statistical variable X (I use upper-case letters 'X', 'Y', 'Z', 'X_1', 'Y_1', 'Z_1', etc. from the end of the alphabet for statistical variables) can assign to individuals in a given domain D are called *values* of X. In the following, '$val(X)$' will stand for the set of all values a variable X can assign to individuals u in a given domain D.

Statistical variables can be discrete or continuous. While $val(X)$ of a *discrete variable* X is finite, the set of possible values $val(X)$ of a *continuous variable* X is infinite. One special kind of discrete variables are binary variables. *Binary variables* are variables with exactly two possible values (1 and 0, *yes* and *no*, or *on* and *off*, etc.). Whenever continuous quantities (e.g., weight, mass, length, etc.) are considered in subsequent chapters, they will be represented by discrete variables sufficiently fine-grained to match the applied measurement methods. This means that $val(X)$ with $|val(X)| = n+1$ of such a variable X will, for example, be identical to $\{[0 \cdot \varepsilon, 1 \cdot \varepsilon],]1 \cdot \varepsilon, 2 \cdot \varepsilon], \ldots,](n-1) \cdot \varepsilon, n \cdot \varepsilon],]n \cdot \varepsilon, \infty[\}$, where ε corresponds to the given measurement accuracy, i.e., ε is the smallest measurable quantity given a certain measurement method. This procedure avoids measure theory and the use of integrals, which makes the subsequent chapters much more accessible.

Since probabilistic statements containing statistical variables can get very complex and convoluted, I will use the following conventions: Whenever reference to specific individuals is not necessary, then (i) formulae like '$P(X(u) = x)$' can be replaced by '$P(X = x)$', while (ii) formulae like '$P(X = x)$' can be replaced by '$P(x)$'. Instead of the quite long '$\forall x \in val(X)$' and '$\exists x \in val(X)$' it is oftentimes more convenient to write '$\forall x$' and '$\exists x$', respectively, for short.

Probability distributions over sets of variables Given a set V of statistical variables $X_1, \ldots, X_n$, $P(X_1, \ldots, X_n)$ is called a *probability distribution* over V if and only if P assigns a value $r_i \in [0, 1]$ to every event $A \in val(X_1) \times \ldots \times val(X_n)$. Given a probability distribution P over $V = \{X_1, \ldots, X_n\}$, all kinds of probabilities can be computed. The probability of the instantiation of a variable X_i to some value x_i, for example, can be computed as $\sum_A P(A)$, where A is an element of $val(X_1) \times \ldots \times val(X_n)$ in which x_i occurs. The probability of the instantiation $x_{i_1}, \ldots, x_{i_m}$ of more than one variable $X_{i_1}, \ldots, X_{i_m} \in V$ is, accordingly, defined as $\sum_A P(A)$, where A is an element of $val(X_1) \times \ldots \times val(X_n)$ in which instantiation $x_{i_1}, \ldots, x_{i_m}$ occurs, etc. In fact, every probability distribution over V gives rise to a probability function P over $\mathcal{P}(val(X_1) \times \ldots \times val(X_n))$, which is the power set algebra over the elementary events $X_1 = x_1, \ldots, X_n = x_n$.

Given a probability distribution $P(X_1, \ldots, X_n)$, for every sequence of statistical variables $X_{i_1}, \ldots, X_{i_m}$ a new statistical variable M can be defined. This can be done in the following way: If we want to introduce a variable M for a sequence of variables $X_{i_1}, \ldots, X_{i_m}$, then the set of possible values of this newly introduced variable M can be defined as $val(M) = val(X_{i_1}) \times \ldots \times val(X_{i_m})$, and the probabilities of M's value instantiations $m = \langle x_{i_1}, \ldots, x_{i_m} \rangle$ are defined as $P(x_{i_1}, \ldots, x_{i_m})$. In the following I will often loosely refer to variables M for sequences of variables $X_{i_1}, \ldots, X_{i_m}$ as a sequence or set of variables.

Whenever a probability distribution P over a variable set $V = \{X_1, \ldots, X_n\}$ is specified, the corresponding probability distribution P' for a subset $V' = \{X_{i_1}, \ldots, X_{i_m}\}$ of V can be defined as $P'(X_{i_1}, \ldots, X_{i_m}) =_{df} P(X_{i_1}, \ldots, X_{i_m})$. So P' coincides with P over the value space $val(X_{i_1}) \times \ldots \times val(X_{i_m})$. P' is called P's *restriction* to V' and is denoted by '$P \uparrow V'$'.

Sometimes it may be convenient to define a probability distribution that is conditionalized on a certain fixed context $M = m$, where a context is a set of variables tied to certain values. We can define such a distribution $P_m(X)$ as $P_m(X) =_{df} P(X|m)$.

Important notions and theorems The basic axioms of probability calculus as well as the equations introduced in Sect. 2.3 do also hold for probability distributions over sets of statistical variables. I will demonstrate this for the following more important equations and begin with the law of total probability. (The given equations can be motivated in the same way as their counterparts in Sect. 2.3.) 'A' and 'B_i' in Equation 2.3 must be specified to 'x' and 'y', respectively, where 'x' stands for a value instantiation of a variable X and 'y' ranges over the possible values of a variable Y. (Note that X and Y may also be sets of variables.)

Now the following equation holds in any probability distribution P:

$$P(x) = \sum_{y \in val(Y)} P(x, y) \tag{2.11}$$

Whenever $P(y) > 0$ holds, the conditional probability of x given y for statistical variables is defined as follows:

$$P(x|y) =_{df} \frac{P(x, y)}{P(y)} \tag{2.12}$$

The product rule for statistical variables:

$$P(x, y) = P(x|y) \cdot P(y) \tag{2.13}$$

The chain rule formula for statistical variables:

$$P(x_1, \ldots, x_n) = P(x_1) \cdot P(x_2|x_1) \cdot \ldots \cdot P(x_n|x_1, \ldots, x_{n-1}) \tag{2.14}$$

Or equivalently:

$$P(x_1, \ldots, x_n) = \prod_{i=1}^{n} P(x_i|x_1, \ldots, x_{i-1}) \tag{2.15}$$

The law of total probability for statistical variables:

$$P(x) = \sum_{y \in val(Y)} P(x|y) \cdot P(y) \tag{2.16}$$

And last but not least, Bayes' theorem for statistical variables: Whenever $P(x) > 0$, then:

$$P(y|x) = \frac{P(x|y) \cdot P(y)}{P(x)} \tag{2.17}$$

Equations 2.11, 2.13, 2.14, 2.15, and 2.16 can be generalized for contexts $M = m$ as follows: If (i) A in the formula $P(A)$ appearing at the right hand side of the '=' in the respective equation does have the form $\ldots|\ldots$, then just add 'm' at the right-hand side of '|'. If (ii) A does not already have the form $\ldots|\ldots$, then add '$|m$' between 'A' and the bracket ')'. Following this procedure, we get the following conditionalized versions of the theorems presented before:

$$P(x|m) = \sum_{y \in val(Y)} P(x, y|m) \tag{2.18}$$

$$P(x, y|m) = P(x|y, m) \cdot P(y|m) \tag{2.19}$$

$$P(x_1, \ldots, x_n|m) = P(x_1|m) \cdot P(x_2|x_1, m) \cdot \ldots \cdot P(x_n|x_1, \ldots, x_{n-1}, m) \tag{2.20}$$

$$P(x_1, \ldots, x_n|m) = \prod_{i=1}^{n} P(x_i|x_1, \ldots, x_{i-1}, m) \tag{2.21}$$

$$P(x|m) = \sum_{y \in val(Y)} P(x|y, m) \cdot P(y|m) \tag{2.22}$$

2.5   Correlation and Probabilistic Independence

Probabilistic dependence/independence relations   Statistical correlation is a relation among statistical variables or sets of variables.[4] Probabilistic dependence can be defined with respect to a given probability distribution in the following way:

Definition 2.1 (probabilistic dependence)   If P is a probability distribution over variable set V and $X, Y, Z \in V$, then: $DEP_P(X, Y|Z) \equiv_{df} \exists x \exists y \exists z (P(x|y, z) \neq P(x|z) \wedge P(y, z) > 0)$.

Read '$DEP_P(X, Y|Z)$' as 'X is probabilistically dependent on Y conditional on Z in P' or as 'X and Y are correlated given Z in P'. We will follow the convention to identify unconditional dependence $DEP(X, Y)$ with dependence given the empty set $DEP(X, Y|Z = \emptyset)$.

Probabilistic independence can be defined as the negation of statistical correlation:

---

[4]In the following, variables X, Y, Z could also be exchanged by sets of variables X, Y, Z and vice versa, where these sets X, Y, Z have to be treated as new variables as explained in Sect. 2.4.

Definition 2.2 (probabilistic independence) If P is a probability distribution over variable set V and $X, Y, Z \in V$, then: $INDEP_P(X, Y|Z) \equiv_{df} \neg DEP_P(X, Y|Z)$, i.e., $\forall x \forall y \forall z(P(x|y, z) = P(x|z) \vee P(y, z) = 0)$ holds.

Again, we identify unconditional independence $INDEP(X, Y)$ as independence given the empty set $INDEP(X, Y|Z = \emptyset)$.

Properties of probabilistic dependence/independence relations The following properties (which are also called *graphoid axioms*) hold for all probability distributions P (Pearl 2000, p. 11; Dawid 1979; Pearl and Paz 1985):

Symmetry: $INDEP_P(X, Y|Z) \Rightarrow INDEP_P(Y, X|Z)$
Decomposition: $INDEP_P(X, \{Y, W\}|Z) \Rightarrow INDEP_P(X, Y|Z)$
Weak union: $INDEP_P(X, \{Y, W\}|Z) \Rightarrow INDEP_P(X, Y|\{Z, W\})$
Contraction: $INDEP_P(X, Y|Z) \wedge INDEP_P(X, W|\{Z, Y\}) \Rightarrow INDEP_P(X, \{Y, W\}|Z)$

Here is an explanation of the four properties if Z is simply the empty set. In that case the axiom of symmetry states that Y does not depend on X when X does not depend on Y. The axiom of decomposition says that whenever X is independent of both Y and W, then it will also be independent of Y alone. The axiom of weak union tells us that conditionalizing on W does not render X and Y dependent if X is independent of both Y and W. The axiom of contraction finally states that X is independent of both Y and W if X is independent of Y and independent of W when conditionalizing on Y.

2.6 Graph Theory

Graphs are tools for representing diverse kinds of systems and relations among parts of these systems. A *graph* G is an ordered pair $\langle V, E \rangle$, where V is a set consisting of any objects. The elements of V are called the *vertices* of the graph. E is a set of so-called *edges*. The edges of a graph are typically lines and/or arrows (possibly having different kinds of heads and tails) that connect two vertices and capture diverse relations among objects in V. One advantage of graphs is that they can represent the structure of a system in a very vivid way: Whenever the domain of objects we are interested in is finite and the relations among these objects we want to represent are binary, then we can draw a corresponding graph. An example: Suppose we are interested in a population M of five people a_1, a_2, a_3, a_4, and a_5 and in the supervisor relation Q. Suppose further that a_1 is a supervisor of a_3, that a_2 is a supervisor of a_3 and a_4, and that a_3 and a_4 are supervisors of a_5. Thus, the structure of the system is $\langle M, Q \rangle$, where $Q = \{\langle a_1, a_3 \rangle, \langle a_2, a_3 \rangle, \langle a_2, a_4 \rangle, \langle a_3, a_5 \rangle, \langle a_4, a_5 \rangle\}$. To draw a graph $G = \langle V, E \rangle$ capturing this structure this graph's vertex set V should be identical to M. In addition, we need to make some conventions about how E and Q are connected. In our example, let us make the following conventions:

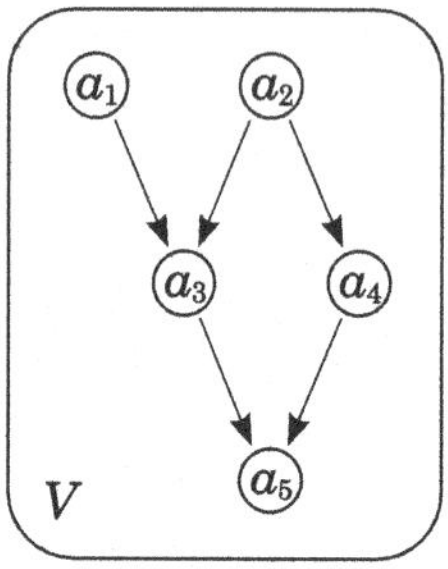

Fig. 2.4 A graph
representing structure $\langle M, Q \rangle$

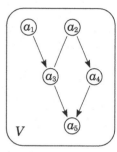

Fig. 2.5 A graph
representing a class of
structures, *viz.*
$\{\langle M, Q \rangle, \langle M, Q' \rangle\}$

- $u \longrightarrow v$ in G if and only if $Q(u, v) \vee Q(v, u)$.
- $u \longrightarrow v$ if and only if $Q(u, v) \wedge \neg Q(v, u)$.

Now we can represent the supervisor relation in population M by graph $G = \langle V, E \rangle$, where $V = M$ and $E = \{a_1 \longrightarrow a_3, a_2 \longrightarrow a_3, a_2 \longrightarrow a_4, a_3 \longrightarrow a_5, a_4 \longrightarrow a_5\}$ (see also Fig. 2.4). According to the two conventions about how E and Q are connected, we can uniquely determine G on the basis of $\langle M, Q \rangle$ as well as $\langle M, Q \rangle$ on the basis of G. Another nice feature of graphs is that they can represent more than one structure—they can capture a whole class of structures. The graph in Fig. 2.5, for instance, represents, according to the conventions made above, $\langle M, Q \rangle$ as well as $\langle M, Q' \rangle$ with $Q' = \{\langle a_1, a_3 \rangle, \langle a_3, a_2 \rangle, \langle a_2, a_4 \rangle, \langle a_3, a_5 \rangle, \langle a_4, a_5 \rangle\}$.[5]

After giving some ideas of what can be done by means of graphs, I will give a brief overview of the graph theoretical terminology relevant for the subsequent chapters: A graph $G = \langle V, E \rangle$ is called a *cyclic graph* if there is at least one chain of edges of the form $u \longrightarrow \ldots \longrightarrow u$ in G; otherwise it is called an *acyclic graph*. A graph $G = \langle V, E \rangle$ is called an *undirected graph* if all edges in E have the form $u \longrightarrow v$. A graph $G = \langle V, E \rangle$ including different kinds of edges (e.g., '$\longrightarrow$' and '$\longrightarrow$') is called a *mixed graph*. A graph $G = \langle V, E \rangle$ whose set of edges E does only contain directed edges ('$\longrightarrow$') is called a *directed graph*. Graphs that are directed as well as acyclic are called *directed acyclic graphs* (or DAGs for short). If V is a set of

---

[5] $\langle M, Q'' \rangle$ with $Q'' = \{\langle a_1, a_3 \rangle, \langle a_2, a_3 \rangle, \langle a_2, a_4 \rangle, \langle a_3, a_2 \rangle, \langle a_3, a_5 \rangle, \langle a_4, a_5 \rangle\}$ is excluded because of the implicit assumption that the supervisor relation is asymmetric.

vertices and all pairs of vertices in V are connected by an edge in graph $G = \langle V, E \rangle$, then G is called a *complete graph* over V.

If two vertices in a graph are connected by an edge, then they are called *adjacent*. A chain π of any kind of edges of the form $u - \ldots - v$ in a graph $G = \langle V, E \rangle$ (where $u, v \in V$) is called a *path* between u and v in G. A path π between u and v that contains a subpath of the form $w_1 \longrightarrow w_2 \longleftarrow w_3$ is called a *collider path* between u and v with w_2 as a *collider* on this path π and is represented by '$u \longrightarrow\longleftarrow v$'. A path π of the form $u \longrightarrow \ldots \longrightarrow v$ is called a *directed path* in G going from u to v and is represented via '$u \longrightarrow\longrightarrow v$'. '$u\overset{\longleftarrow}{\longrightarrow}v$' stands short for a *direct cycle* π of the form $u \longrightarrow v \longrightarrow u$, while '$u\overset{\longleftarrow}{\longrightarrow}\overset{\longleftarrow}{\longrightarrow}v$' is short for a (direct or indirect) *cycle* π of the form $u \longrightarrow\longrightarrow v \longrightarrow\longrightarrow u$.

If the set of edges of a graph contains one or more arrows, the following family-terminology can be used for describing several relations among objects in G's vertice set V: Whenever u and v are connected by a directed path π ($u \longrightarrow\longrightarrow v$) in $G = \langle V, E \rangle$, then u is called an *ancestor* of v in G and v is called a *descendant* of u in G. The set of all ancestors of a vertex u shall be referred to via '$Anc_{\langle V,E \rangle}(u)$', while '$Des_{\langle V,E \rangle}(u)$' is used to designate the set of descendants of u. A path π between u and v of the form $u \longleftarrow\longleftarrow w \longrightarrow\longrightarrow v$ such that no vertex on π appears more often than once on π is called a *common ancestor path* between u and v (with w as a *common ancestor* of u and v) and is represented by '$u \longleftarrow\longrightarrow v$'. Whenever $u \longrightarrow v$ holds in G, then u is called a *parent* of v, while v is called a *child* of u. '$Par_{\langle V,E \rangle}(u)$' shall stand for the set of parents of u while '$Chi_{\langle V,E \rangle}(u)$' shall refer to the set of children of u in graph G.

2.7 Bayesian Networks

Markovian parents Bayesian networks were originally developed to compactly represent probability distributions and to simplify probabilistic reasoning (Neapolitan 1990; Pearl 1988). The main idea behind the concept of a Bayesian network is the following: As seen in Sect. 2.4, a probability distribution P over a set of variables V is specified by assigning a value $r_i \in [0, 1]$ to every instantiation $V = v$. Since $|val(V)|$ becomes horribly large even if V contains only a few variables, a lot of space would be required to write the whole probability distribution down, while computing probabilities for specific events $M = m$ (with $M \subseteq V$) can consume a lot of time and resources. So is there any possibility to store probability distributions in a more compact way? The formalism of Bayesian networks provides a positive answer to this question (provided the corresponding graph is sparse; for details see below). According to the chain rule formula (Equation 2.14), $P(x_1, \ldots, x_n) = \prod_{i=1}^{n} P(x_i | x_1, \ldots, x_{i-1})$ holds for arbitrary value instantiations $x_1, \ldots, x_n$ of arbitrary orderings of variables $X_1, \ldots, X_n \in V$, and thus, we can specify a probability distribution also by assigning values $r_i \in [0, 1]$ to every possible value x_i of every variable $X_i \in V$ conditional on all possible combinations of values $x_1, \ldots, x_{i-1}$ of X_i's predecessors $X_1, \ldots, X_{i-1}$ in the given ordering.

Table 2.1 Exemplary probability distribution P over $V = \{X, Y, Z\}$

$$P(x_1, y_1, z_1) = \frac{9}{32}$$
$$P(x_1, y_1, z_2) = \frac{3}{32}$$
$$P(x_1, y_2, z_1) = \frac{2}{32}$$
$$P(x_1, y_2, z_2) = \frac{2}{32}$$
$$P(x_2, y_1, z_1) = \frac{6}{32}$$
$$P(x_2, y_1, z_2) = \frac{2}{32}$$
$$P(x_2, y_2, z_1) = \frac{4}{32}$$
$$P(x_2, y_2, z_2) = \frac{4}{32}$$

Table 2.2 Application of the chain rule formula to P

$$P(x_1, y_1, z_1) = P(x_1) \cdot P(y_1|x_1) \cdot P(z_1|x_1, y_1) = \frac{9}{32}$$
$$P(x_1, y_1, z_2) = P(x_1) \cdot P(y_1|x_1) \cdot P(z_2|x_1, y_1) = \frac{3}{32}$$
$$P(x_1, y_2, z_1) = P(x_1) \cdot P(y_2|x_1) \cdot P(z_1|x_1, y_2) = \frac{2}{32}$$
$$P(x_1, y_2, z_2) = P(x_1) \cdot P(y_2|x_1) \cdot P(z_2|x_1, y_2) = \frac{2}{32}$$
$$P(x_2, y_1, z_1) = P(x_2) \cdot P(y_1|x_2) \cdot P(z_1|x_2, y_1) = \frac{6}{32}$$
$$P(x_2, y_1, z_2) = P(x_2) \cdot P(y_1|x_2) \cdot P(z_2|x_2, y_1) = \frac{2}{32}$$
$$P(x_2, y_2, z_1) = P(x_2) \cdot P(y_2|x_2) \cdot P(z_1|x_2, y_2) = \frac{4}{32}$$
$$P(x_2, y_2, z_2) = P(x_2) \cdot P(y_2|x_2) \cdot P(z_2|x_2, y_2) = \frac{4}{32}$$

Table 2.3 P can also be specified by the conditional probabilities above

$P(x_1) = \frac{1}{2}$	$P(y_1|x_1) = \frac{3}{4}$	$P(z_1|x_1, y_1) = \frac{3}{4}$
	$P(y_1|x_2) = \frac{1}{2}$	$P(z_1|x_1, y_2) = \frac{1}{2}$
		$P(z_1|x_2, y_1) = \frac{3}{4}$
		$P(z_1|x_2, y_2) = \frac{1}{2}$

Here is an example demonstrating how this procedure can be used to store probability distributions in a more compact way: Assume X, Y, and Z are binary variables with $val(X) = \{x_1, x_2\}$, $val(Y) = \{y_1, y_2\}$, and $val(Z) = \{z_1, z_2\}$. Assume further that P is a probability distribution over $V = \{X, Y, Z\}$ determined by the equations in Table 2.1. Here we need eight equations, one for each elementary event, to specify P. Given the ordering X, Y, Z, the equations in Table 2.2 hold due to the chain rule formula (Equation 2.14). It follows that P can also be specified by the factors appearing in the equations in Table 2.2, i.e., by the seven equations in Table 2.3.

We can write down P in an even more compact way. For this purpose, the notion of the set of the *Markovian parents* (Par^M) of a variable X_i in a given ordering $X_1, \ldots, X_n$ will be helpful (cf. Pearl 2000, p. 14):

Definition 2.3 (Markovian parents) If P is a probability distribution over variable set V and $X_1, \ldots, X_n$ is an ordering of the variables in V, then for all $X_i \in V$ and

$M \subseteq V$: M is the set of Markovian parents of X_i if and only if M is the narrowest subset of $\{X_1, \ldots, X_{i-1}\}$ for which $INDEP_P(X_i, \{X_1, \ldots, X_{i-1}\}|M)$ holds.

In other words, the set of Markovian parents of a variable X_i in a given ordering $X_1, \ldots, X_n$ is the narrowest subset of $\{X_1, \ldots, X_{i-1}\}$ such that conditionalizing on $Par^M(X_i)$ makes X_i probabilistically independent of all its predecessors $X_1, \ldots, X_{i-1}$ in this ordering.

Let us now have a closer look at how specifying P in the example discussed above can be simplified by Definition 2.3: Since there are no predecessors of X, $Par^M(X)$ is the empty set. $Par^M(Y)$ cannot be the empty set since $P(y_1) = P(y_1|x_1) \cdot P(x_1) + P(y_1|x_2) \cdot P(x_2) = \frac{3}{4} \cdot \frac{1}{2} + \frac{1}{2} \cdot \frac{1}{2} = \frac{5}{8} \neq \frac{3}{4} = P(y_1|x_1)$ holds, and thus, $DEP_P(X, Y)$. It follows trivially from the definition of conditional probabilistic dependence (Definition 2.2) that Y is probabilistically independent of $\{X\}$ conditional on $\{X\}$, and thus, $Par^M(Y)$ must be $\{X\}$. Since the following equations hold due to Equations 2.16 and 2.17, we get $INDEP_P(Z, \{X, Y\}|\{Y\})$ with Definition 2.2, and hence, $\{Y\}$ seems to be a good candidate for the set of Markovian parents of Z:

$$P(z_1|y_1) = P(z_1|x_1, y_1) \cdot P(x_1|y_1) + P(z_1|x_2, y_1) \cdot P(x_2|y_2)$$

$$= P(z_1|x_1, y_1) \cdot \frac{P(y_1|x_1) \cdot P(x_1)}{P(y_1)} + P(z_1|x_2, y_1) \cdot \frac{P(y_1|x_2) \cdot P(x_2)}{P(y_1)}$$

$$= P(z_1|x_1, y_1) \cdot \frac{P(y_1|x_1) \cdot P(x_1)}{P(y_1|x_1) \cdot P(x_1) + P(y_1|x_2) \cdot P(x_2)}$$

$$+ P(z_1|x_2, y_1) \cdot \frac{P(y_1|x_2) \cdot P(x_2)}{P(y_1|x_1) \cdot P(x_1) + P(y_1|x_2) \cdot P(x_2)}$$

$$= \frac{3}{4} \cdot \frac{\frac{3}{4} \cdot \frac{1}{2}}{\frac{3}{4} \cdot \frac{1}{2} + \frac{1}{2} \cdot \frac{1}{2}} + \frac{3}{4} \cdot \frac{\frac{1}{2} \cdot \frac{1}{2}}{\frac{3}{4} \cdot \frac{1}{2} + \frac{1}{2} \cdot \frac{1}{2}} = \frac{3}{4} = P(z_1|x_1, y_1) = P(z_1|x_2, y_1)$$

$$(2.23)$$

$$P(z_1|y_2) = P(z_1|x_1, y_2) \cdot P(x_1|y_2) + P(z_1|x_2, y_2) \cdot P(x_2|y_2)$$

$$= P(z_1|x_1, y_2) \cdot \frac{P(y_2|x_1) \cdot P(x_1)}{P(y_2)} + P(z_1|x_2, y_2) \cdot \frac{P(y_2|x_2) \cdot P(x_2)}{P(y_2)}$$

$$= P(z_1|x_1, y_2) \cdot \frac{P(y_2|x_1) \cdot P(x_1)}{P(y_2|x_1) \cdot P(x_1) + P(y_2|x_2) \cdot P(x_2)}$$

$$+ P(z_1|x_2, y_2) \cdot \frac{P(y_2|x_2) \cdot P(x_2)}{P(y_2|x_1) \cdot P(x_1) + P(y_2|x_2) \cdot P(x_2)}$$

$$= \frac{1}{2} \cdot \frac{\frac{1}{4} \cdot \frac{1}{2}}{\frac{1}{4} \cdot \frac{1}{2} + \frac{1}{2} \cdot \frac{1}{2}} + \frac{1}{2} \cdot \frac{\frac{1}{2} \cdot \frac{1}{2}}{\frac{1}{4} \cdot \frac{1}{2} + \frac{1}{2} \cdot \frac{1}{2}} = \frac{1}{2} = P(z_1|x_1, y_2) = P(z_1|x_2, y_2)$$

$$(2.24)$$

Table 2.4 The concept of a variable's Markovian parents allows for an even more compact specification of P

$P(x_1) = \frac{1}{2}$	$P(y_1\|x_1) = \frac{3}{4}$	$P(z_1\|y_1) = \frac{3}{4}$
	$P(y_1\|x_2) = \frac{1}{2}$	$P(z_1\|y_2) = \frac{1}{2}$

With help of these equations it is easy to see that $P(z_1) = P(z_1|y_1) \cdot P(y_1) + P(z_1|y_2) \cdot P(y_2) = \frac{3}{4} \cdot \frac{5}{8} + \frac{1}{2} \cdot \frac{3}{8} = \frac{11}{32} \neq \frac{3}{4} = P(z_1|y_1)$, and thus, also $DEP_P(Y, Z)$ holds. So $\emptyset$ cannot be the set of Markovian parents of Z. Therefore, $\{Y\}$ is in fact the narrowest set of predecessors of Z in the given ordering X, Y, Z that screens Z off from all its predecessors, and hence, $\{Y\}$ is the much sought-after set of Markovian parents of Z. So we can store our exemplary probability distribution P via the five (instead of the original eight) equations in Table 2.4 determining the probabilities of the diverse variable values conditional on their Markovian parents.

Summarizing the considerations above, it turns out that the following equation holds for any given ordering of variables $X_1, \ldots, X_n$—this equation provides a simplification of the chain rule formula Equation 2.15 for fixed orderings by means of the notion of a variable's Markovian parents[6]:

$$P(x_1, \ldots, x_n) = \prod_{i=1}^{n} P(x_i|x_1, \ldots, x_{i-1}) = \prod_{i=1}^{n} P(x_i|par^M(X_i)) \qquad (2.25)$$

Markov condition and Markov compatibility Let us come to the notion of a Bayesian network now. Bayesian networks are closely connected to the notion of Markovian parents. A Bayesian network combines a probability distribution P over a set of variables V with a graph over V in such a way that a set of probabilistic independencies that hold in P can be read off this graph's structure. A *Bayesian network* (BN) is an ordered pair $\langle G, P \rangle$, where $G = \langle V, E \rangle$ is a DAG (whose vertex set contains only statistical variables) and P is a probability distribution over this graph's vertex set V that satisfies the so-called *Markov condition* (MC). A DAG and a probability distribution satisfying MC are also said to be *Markov compatible* (cf. Pearl 2000, p. 16).

Definition 2.4 (Markov condition) A graph $G = \langle V, E \rangle$ and a probability distribution P over V satisfy the Markov condition if and only if it holds for all $X \in V$ that $INDEP_P(X, V \backslash Des_G(X)|Par_G(X))$.

So $\langle G, P \rangle$ is a BN if and only if all variables X in G's vertex set V are probabilistically independent of all non-descendants of X conditional on X's parents. Since BNs satisfy the Markov condition, MC and the BN whose graph is depicted

---

[6]While '$Par^M(X)$' denotes the set of X's Markovian parents, '$par^M(X)$' stands for the instantiation of X's Markovian parents $Par^M(X)$ induced by $x_1, \ldots, x_n$ in $P(x_1, \ldots, x_n)$ on the left hand side of Equation 2.25.

Fig. 2.6 Exemplary DAG

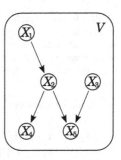

Table 2.5 Independencies
implied by MC and the graph
depicted in Fig. 2.6

$INDEP_P(X_1, \{X_3\}|\emptyset)$
$INDEP_P(X_2, \{X_3\}|\{X_1\})$
$INDEP_P(X_3, \{X_1, X_2, X_4\}|\emptyset)$
$INDEP_P(X_4, \{X_1, X_3, X_5\}|\{X_2\})$
$INDEP_P(X_5, \{X_1, X_4\}|\{X_2, X_3\})$

in Fig. 2.6 imply, for instance, the probabilistic independence relations in Table 2.5
for the associated probability distribution P over V. (The independencies following
trivially from Definition 2.2, e.g., $INDEP(X_1, \{X_1\}|\emptyset)$ or $INDEP(X_2, \{X_1\}|\{X_1\})$, are
not mentioned in this list.)

If an ordering $X_1, \ldots, X_n$ of variables in V corresponds to the ordering of these
variables in a BN $\langle V, E, P \rangle$ (this means that there is no arrow '$\longrightarrow$' in the BN's graph
$G = \langle V, E \rangle$ pointing from a variables X_j to a variable X_i, where X_i is a predecessor
of X_j in the given ordering), then the set of Markovian parents of every variable in
this ordering is a subset of this variable's set of parents[7]:

$$Par^M(X) \subseteq Par_{\langle V,E \rangle}(X) \tag{2.26}$$

$$P(x_1, \ldots, x_n) = \prod_{i=1}^{n} P(x_i|x_1, \ldots, x_{i-1}) = \prod_{i=1}^{n} P(x_i|par^M(X_i))$$

$$= \prod_{i=1}^{n} P(x_i|par_{\langle V,E \rangle}(X_i)) \tag{2.27}$$

Minimality condition The minimality condition can be defined as follows:

Definition 2.5 (minimality condition) A graph $G = \langle V, E \rangle$ and a probability
distribution P over V satisfy the minimality condition if and only if $\langle V, E, P \rangle$
satisfies MC and there is no submodel $\langle V, E', P \rangle$ of $\langle V, E, P \rangle$ with $E' \subset E$ that
satisfies MC.

---

[7]Here is an example where the inclusion in Equation 2.26 is strict: Assume our BN has the graph
depicted in Fig. 2.6. Assume further that X_5 depends on X_2, but not on X_3. In that case, X_5's only
Markov parent in the ordering $X_1, \ldots X_5$ is X_2, while X_5's graphical parents are X_2 and X_3.

Table 2.6 If the graph depicted in Fig. 2.6 is interpreted as the graph of a minimal BN, then the independence and dependence relations below are implied by this BN's topological structure; if interpreted as the graph of a non-minimal BN, on the other hand, only the independencies in the left column are implied. The 'M' on the right hand side of the stroke '$|$' functions as a proxy for some subset of $V = \{X_1, \ldots X_5\}$ not containing the variables X_i and X_j on the left hand side of the '$|$'

Independence relations	Dependence relations
$INDEP_P(X_1, \{X_3\}|\emptyset)$	$DEP_P(X_1, \{X_2\}|M)$
$INDEP_P(X_2, \{X_3\}|\{X_2\})$	$DEP_P(X_2, \{X_4\}|M)$
$INDEP_P(X_3, \{X_1, X_2, X_4\}|\emptyset)$	$DEP_P(X_2, \{X_5\}|M)$
$INDEP_P(X_4, \{X_1, X_3, X_5\}|\{X_2\})$	$DEP_P(X_3, \{X_5\}|M)$
$INDEP_P(X_5, \{X_1, X_4\}|\{X_2, X_3\})$	

A BN that satisfies the minimality condition (cf. Spirtes et al. 2000, p. 12) is called a *minimal Bayesian network*. In a minimal BN, every connection between two variables by an arrow $X \longrightarrow Y$ is probabilistically productive. So the graph $\langle V, E \rangle$ of a minimal BN $\langle V, E, P \rangle$ does not only imply some probabilistic independence relations, but also some probabilistic dependence relations that have to hold in any compatible probability distribution P. If the graph depicted in Fig. 2.6, for instance, is the graph of a minimal BN $\langle V, E, P \rangle$, then the dependence/independence relations in Table 2.6 have to hold in P. (Also all dependence and independence relations implied by the relations in Table 2.6 have, of course, to hold in P.)

If a BN $\langle V, E, P \rangle$ does satisyfy the minimality condition, then also the following stronger version of Equation 2.26 holds:

$$Par^M(X) = Par_{\langle V, E \rangle}(X) \qquad (2.28)$$

***d*-separation and *d*-connection** If we take a look at any arbitrarily chosen directed acyclic graph $G = \langle V, E \rangle$, then, thanks to MC, we can read off the graph which probabilistic independence relations have to hold in any probability distribution P Markov-compatible with G. The graph depicted in Fig. 2.7, for instance, is compatible with any probability distribution P including the probabilistic independence relations in Table 2.7. So we know a lot about a BN's probability distribution only by looking at the topology of this BN's graph. But is there a way we can learn even more from a BN's graph? MC does often not tell us directly whether two variables of a BN's graph are probabilistically dependent/independent and how their probabilistic dependence/independence would be affected when one conditionalizes on certain variables or sets of variables of this graph. So, for instance, are X_2 and X_4 correlated conditional on X_3 in Fig. 2.7? We could try to use the independencies in Table 2.7 together with the graphoid axioms introduced in Sect. 2.5 to answer this question without tedious probabilistic computation. But maybe there is an even simpler way to answer the question just by looking at the BN's graph?

Fortunately, the answer to this question is an affirmative one: There is a strong connection between the topological properties of a BN's DAG and the diverse

Fig. 2.7 Another exemplary DAG

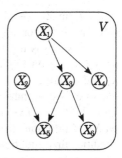

Table 2.7 Independencies implied by MC and the graph depicted in Fig. 2.7

$INDEP_P(X_1, \{X_2\}|\emptyset)$

$INDEP_P(X_2, \{X_1, X_3, X_4, X_6\}|\emptyset)$

$INDEP_P(X_3, \{X_2, X_4\}|\{X_1\})$

$INDEP_P(X_4, \{X_2, X_3, X_5, X_6\}|\{X_1\})$

$INDEP_P(X_5, \{X_1, X_4, X_6\}|\{X_2, X_3\})$

$INDEP_P(X_6, \{X_1, X_2, X_4, X_5\}|\{X_3\})$

dependence/independence relations which may or may not hold in associated probability distributions P. This connection is established via the notion of *d-separation/d-connection* (cf. Pearl 2000, pp. 16f)[8]:

Definition 2.6 (*d*-separation/*d*-connection) $X, Y \in V$ are *d*-separated by $M \subseteq V\setminus\{X, Y\}$ in directed graph $G = \langle V, E\rangle$ ($SEP^d_{\langle V,E\rangle}(X, Y|M)$) if and only if every path π between X and Y contains a subpath of one of the following forms (where $Z_1, Z_2, Z_3 \in V$):

(a) $Z_1 \longrightarrow Z_2 \longrightarrow Z_3$ with $Z_2 \in M$, or
(b) $Z_1 \longleftarrow Z_2 \longrightarrow Z_3$ with $Z_2 \in M$, or
(c) $Z_1 \longrightarrow Z_2 \longleftarrow Z_3$ with $Z_2 \notin M$, where also no descendant of Z_2 is in M.

X and Y are *d*-connected given M in $G = \langle V, E\rangle$ ($CON^d_{\langle V,E\rangle}(X, Y|M)$) if and only if they are not *d*-separated by M.

When there is a path π between X and Y that goes through M, then we say that M *blocks* this path π if M *d*-separates X and Y, i.e., $SEP^d_{\langle V,E\rangle}(X, Y|M)$ holds. A path π not blocked by M is said to be *activated* by M.

Here are some examples for illustrating how Definition 2.6 works: In the graph in Fig. 2.7, X_1 and X_3 are *d*-connected (given the empty set) because there is a path between X_1 and X_3, *viz.* $X_1 \longrightarrow X_3$, not satisfying one of the conditions (a)–(c) in Definition 2.6. X_4 and X_5 are also *d*-connected (given the empty set); they are connected via path $\pi: X_5 \longleftarrow X_3 \longleftarrow X_1 \longrightarrow X_4$ which contains no subpath of the form $Z_1 \longrightarrow Z_2 \longrightarrow Z_3$ with $Z_2 \in \emptyset$ (thus, (a) is not satisfied), no subpath

---

[8] The term '*d*-connection' is due to the fact that *d*-connection was initially defined for directed graphs; hence, the '*d*' for 'directed'.

of the form $Z_1 \longleftarrow Z_2 \longrightarrow Z_3$ with $Z_2 \in \emptyset$ (thus, (b) is not satisfied), and no subpath of the form $Z_1 \longrightarrow Z_2 \longleftarrow Z_3$ with $Z_2 \notin \emptyset$ and no descendant of Z_2 in $\emptyset$ (thus, (c) is not satisfied). This path π is blocked when one conditionalizes on X_1, X_3, or on $\{X_1, X_3\}$. X_1 and X_2 are d-separated (by the empty set) because X_1 and X_2 are connected only by a collider path where neither the collider X_5 nor one of X_5's descendants (actually, there are none) is an element of the empty set. Though $SEP^d_{\langle V,E \rangle}(X_1, X_2 | \emptyset)$, X_1 and X_2 become d-connected when conditionalizing on X_5. (Conditions (a)–(c) in Definition 2.6 are not satisfied in that case.) If one conditionalizes not only on X_5, but also on X_3 (i.e., on the set $\{X_3, X_5\}$), probability propagation over the path $X_2 \longrightarrow X_5 \longleftarrow X_3 \longleftarrow X_1 \longrightarrow X_4$ is blocked again, because condition (a) of Definition 2.6 is satisfied in that case.

d-separation/d-connection and probabilistic dependence/independence relations of a BN's probability distribution P are connected via the *d-separation criterion* (cf. Pearl 2000, p. 18):

Criterion 2.1 (*d*-separation criterion) *If graph* $G = \langle V, E \rangle$ *and probability distribution* P *over* V *satisfy MC, then* $INDEP_P(X, Y | M)$ *holds for all* $X, Y \in V$ *and* $M \subseteq V \setminus \{X, Y\}$ *whenever* $SEP^d_{\langle V,E \rangle}(X, Y | M)$ *holds.*

The d-separation criterion identifies all and only the independencies also implied by MC (cf. Pearl 2000, p. 19). Criterion 2.1 finally allows one to read off the independence relations which have to hold in any probability distribution P compatible to a given directed graph G. X_4 and X_5 are, for instance, d-connected in the graph depicted in Fig. 2.7, and thus, X_4 and X_5 may be correlated in the BN's associated probability distribution. (It may also be the case that X_4 and X_5 are independent.) When one conditionalizes on any set of variables lying on the path $\pi \colon X_5 \longleftarrow X_3 \longleftarrow X_1 \longrightarrow X_4$, e.g., on X_1, then this path is blocked, and thus, Criterion 2.1 implies $INDEP_P(X_5, X_4 | X_1)$.

Chapter 3
Causal Nets

Abstract In this chapter I give a brief overview of the central notions and definitions of the causal nets framework, which was developed in detail by Pearl and Spirtes, Glymour, and Scheines. I also begin to develop and justify an axiomatization of the causal nets framework, to be continued and finished in Chap. 4.

3.1 Introduction

In this chapter I give a brief overview of the central notions and definitions of the causal nets framework, which has its origins in the Bayesian network formalism (Neapolitan 1990; Pearl 1988) and was developed in detail by Pearl (2000) and Spirtes et al. (2000). I also begin to develop and justify an axiomatization of the approach, to be continued and finished in Chap. 4. The causal Bayes net approach, especially in Spirtes et al.'s (2000) understanding, has many in common with modern empirical theories of the sciences and is clearly inspired by the success of the empirical sciences. It treats causation as a primitive relation that is not directly observable, but connected to empirically accessible concepts in certain ways given certain conditions are satisfied. It hence understands causation, like Newtonian physics does w.r.t. force, as a theoretical concept (cf. Glymour 2004; Schurz and Gebharter 2016).

The causal Bayes net framework can be traced back to Wright (1921) and Blalock (1961). The basic ideas underlying the approach are also explicit in Reichenbach (1935/1971, 1956/1991). The idea is that certain causal structures produce specific probabilistic properties by whose means these causal structures can (in principle) be identified. Two of Reichenbach's core ideas were that "if A causes C only through the mediation of a set of variables B, then A and C are statistically independent conditional on B" (Glymour et al. 1991, p. 151) and that "if A does not cause B and B does not cause A, and A and B are statistically dependent, then there exists a common cause of A and B" (ibid.). The causal Bayes net approach picks up these ideas, elaborates them in more detail, and generalizes them to what is nowadays known under the name of the *causal Markov condition* (cf. Williamson 2009). But contrary to Reichenbach, who tried to reduce causation to probability, the causal nets approach is, as already mentioned, much more inspired by modern scientific theories. It abandoned trying to provide an explicit definition of causation, and

© Springer International Publishing AG 2017
A. Gebharter, *Causal Nets, Interventionism, and Mechanisms*,
Synthese Library 381, DOI 10.1007/978-3-319-49908-6_3

rather aims for characterizing causation implicitly by means of the causal Markov condition and several additional conditions. Or, in Glymour's (2004) words: The approach is not a "Socratic", but a "Euclidic" one.

This chapter is structured as follows: I start in Sect. 3.2 by introducing the notions of causal graph and causal model together with some terminological conventions. I proceed by presenting and explaining the already mentioned causal Markov condition in Sect. 3.3 as well as an equivalent and philosophically more transparent formulation, *viz.* the *d*-connection condition. In Sect. 3.4 I discuss two additional assumptions typically needed for causal discovery, *viz.* the causal minimality condition and the so-called causal faithfulness condition, and how they connect to the causal Markov condition. I will also present and discuss alternative and philosophically more transparent and often more useful formulations of these two conditions. Note that all of the mentioned conditions were originally formulated as conditions in (Spirtes et al. 2000) which may or may not be satisfied by systems in the world. I propose and justify corresponding axioms for the causal minimality condition and for a weaker version of the causal faithfulness condition in this chapter. In the next chapter (i.e., Chap. 4), I develop and justify also an axiom corresponding to the causal Markov condition. The causal Bayes net framework is often seen as a purely methodological tool by philosophers. The result of the axiomatization developed in this and the next chapter is intended to show that one can easily construct a full-fledged theory of causation out of this framework.

3.2   Causal Graphs and Causal Models

The causal nets approach uses graph theoretical notions to represent causal structures. A graph $G = \langle V, E \rangle$ is called a *causal graph* (or a causal structure) if it is a directed graph whose arrows ('$\longrightarrow$') are interpreted as direct causal dependencies relative to the graph's variable set V, where the variable at the arrow's tail is called a *direct cause* of the variable at the arrow's head in the respective causal graph.

The family terminology used for relationships between vertices of directed acyclic graphs (DAGs) introduced in Sect. 2.6 has a one-to-one translation to a terminology describing certain causal relationships between variables in causal structures: The parents of a variable X in a causal graph $G = \langle V, E \rangle$, i.e., the elements of $Par_G(X)$, are X's *direct causes* in G, while X's children (the elements of $Chi_G(X)$) are called X's *direct effects* in G. A variable X's ancestors in a given causal graph (the elements of $Anc_G(X)$) are interpreted as X's (direct or indirect) *causes*, and the descendants of X (the elements of $Des_G(X)$) are called X's (direct or indirect) *effects* in G.

A path π in a causal graph is called a *causal path*. A path of the form $X \longrightarrow \longrightarrow Y$ is called a *directed causal path* from X to Y. If Z is a vertex lying somewhere between X and Y on a directed causal path $X \longrightarrow \longrightarrow Y$, then Z is called an *intermediate cause* lying on this path. A path of the form $X \longleftarrow \longleftarrow Z \longrightarrow \longrightarrow Y$ is called a *common cause path* between X and Y (with Z as a common cause of X and

Y lying on this path) if no variable appears more often than once on the path. A path of the form $X \longrightarrow \longrightarrow Z \longleftarrow \longleftarrow Y$ is called a *common effect path* connecting X and Y (with Z as a common effect of X and Y lying on this path). A direct cyclic path $X \overset{\longleftarrow}{\longrightarrow} Y$ is called a *direct causal cycle*, and a (direct or indirect) cyclic path $X \overset{\longleftarrow}{\longrightarrow} \overset{\longleftarrow}{\longrightarrow} Y$ a (direct or indirect) *causal cycle*.

If one supplements a causal graph $G = \langle V, E \rangle$ by a probability distribution P over this causal graph's variable set V, then the resulting triple $S = \langle V, E, P \rangle$ is called a *causal model* (or a causal system). While the graph G represents a causal structure, P is intended to provide information about the strenghts of the causal influences propagated along the causal paths of this structure. The idea is that causation is something not directly observable but ontologically real that produces certain probabilistic phenomena by whose means the causal structures underlying these phenomena can (at least in principle) be identified.

Before we can go on, two important notions closely related to causal models have to be introduced: (i) the notion of a causal model's parameters and (ii) the notion of a causal background context. (i) A causal model's *parameters* are defined as the conditional probabilities $P(x_i | par(X_i))$ for all X_i in the causal model's variable set $V = \{X_1, \ldots, X_n\}$. The parameters $P(x_i | par(X_i))$ of a variable X_i correspond to the causal strengths of the influence of instantions of X_i's parents $Par(X_i) = par(X_i)$ on $X_i = x_i$.

(ii) When we take a look at certain causal models $S = \langle V, E, P \rangle$, there will typically be a fixed causal background context $C = c$. By such a *causal background context* I mean a set C of variables not included in the model of interest's variable set V whose values do not vary. Note that changing a given background context $C = c$ to $C = c'$ (with $c \neq c'$) typically will change the model of interest's associated probability distribution P. Sometimes changing the context might also make it necessary to change the topology of a model's graph. If not explicitly stated, we will by default assume that the background contexts of the systems we are interested in do not vary.

3.3   The Causal Markov Condition

In this section I introduce the causal Bayes net framework's core condition, the so-called *causal Markov condition* (CMC). When satisfied, it connects causal structures, which describe (or model) unobservable causal relationships, to empirically accessible probability distributions (cf. Spirtes et al. 2000, p. 29):

Condition 3.1 (causal Markov condition) *A causal model $S = \langle V, E, P \rangle$ satisfies the causal Markov condition if and only if every $X \in V$ is probabilistically independent of its non-effects conditional on its direct causes.*

The causal Markov condition is essentially the causal interpretation of the Markov condition (see Sect. 2.7) for Bayesian networks. It works exactly like the Markov condition in Bayesian networks and is intended to hold only for causal systems whose associated graphs are DAGs: Given a causal graph, the causal Markov condition implies a set of probabilistic independence relations that have to hold in any compatible probability distribution. Given a causal model, which may represent a complex causal hypothesis, it hence excludes certain logically possible probability distributions over this causal model's variable set. So causal models (or hypotheses) generate predictions; they are empirically testable.

Note that the causal Markov condition (CMC) is formulated as a condition, and not as an axiom. It is not supposed to hold for all causal models. It will be regularly violated in systems whose variable sets V do not contain every common cause of every pair of variables in V. CMC can, however, be expected to hold in causal systems whose variable sets V contain every such common cause. Such variable sets V are called *causally sufficient* (cf. Spirtes et al. 2000, p. 22). If our system of interest is not causally sufficient or we do not know whether it is causally sufficient, it is typically assumed that CMC holds for some expansion of the system (cf. Spirtes et al. 2000, pp. 124f).

In causal models whose graphs are DAGs, such as causally interpreted BNs (see Sect. 2.7), for example, all (and only the) independence relations implied by the causal Markov condition can be identified by applying the *d*-separation criterion (Criterion 2.1).[1] Because of this, the following condition, which is called the *d-connection condition* in Schurz and Gebharter (2016, p. 1084), is equivalent to the causal Markov condition:

Condition 3.2 (*d*-connection condition) *Graph* $G = \langle V, E \rangle$ *and probability distribution* P *over* V *satisfy the d-connection condition if and only if* $CON_G^d(X, Y|M)$ *holds for all* $X, Y \in V$ *and* $M \subseteq V \backslash \{X, Y\}$ *whenever* $DEP_P(X, Y|M)$ *holds.*

The equivalence between the *d*-connection condition and the causal Markov condition reveals the full content of the causal Markov condition: It explicitly states that whenever we have found a correlation between two variables X and Y (relative to a subset M of the system of interest), then this correlation is brought about by some causal connection between these two variables (given M). Correlations are produced by causal connections; they can be *explained* by referring to the relevant parts of the system's underlying causal structure. So what causal relations do is to produce correlations; they explain regularities or the absence of regularities (in case of *d*-separation).

One important difference between CMC and the *d*-connection condition is that the former is assumed to hold only for causal models whose graphs are DAGs, while the latter is assumed to also hold for causal systems whose associated causal structures are directed graphs featuring causal cycles. When the causal model of interest's graph is a directed cyclic graph, then CMC and the *d*-connection condition will typically imply different probabilistic independence relations for the model's associated probability distribution P. Let me briefly illustrate this on the graph depicted in Fig. 3.1: According to the causal Markov condition, every X_i in this

---

[1] For a proof see Lauritzen et al. (1990), Pearl (1988, p. 119f), or Verma (1987).

Fig. 3.1 CMC applied to the causal graph depicted above implies $INDEP_P(X_1, X_4)$, $INDEP_P(X_2, X_4 | \{X_1, X_3\})$, and $INDEP_P(X_3, X_1 | \{X_2, X_4\})$, while the d-connection condition only implies one of these three independence relations, *viz.* $INDEP_P(X_1, X_4)$

graph has to be probabilistically independent of its non-effects conditional on its direct causes. So CMC implies that X_1 has to be probabilistically independent of X_4, that X_2 has to be independent of X_4 given its direct causes X_1 and X_3, and that X_3 has to be probabilistically independent of X_1 when conditionalizing on $\{X_2, X_4\}$. The d-connection condition, on the other hand, entails that two variables X_i and X_j (with $i \neq j$) of the graph are probabilistically independent conditional on a (maybe empty) set M of variables not containing X_i and X_j if X_i and X_j are causally separated by M. In our graph depicted in Fig. 3.1 X_1 and X_4 are d-separated (by the empty set), X_2 and X_4 are d-connected given $\{X_1, X_3\}$, and X_3 and X_1 are d-connected given $\{X_2, X_4\}$. So the d-connection condition implies only one of the three probabilistic independence relations implied by CMC, *viz.* $INDEP_P(X_1, X_4)$.

The d-connection condition is preferable to CMC because it does—contrary to CMC—fit the idea that blocking all paths between two variables should render them independent. Conditionalizing on $\{X_1, X_3\}$, for example, blocks path $X_1 \longrightarrow X_2 \longrightarrow X_3$, but at the same time, activates $X_1 \longrightarrow X_2 \longleftarrow X_3$. So it seems plausible that X_2 and X_4 can be dependent conditional on $\{X_1, X_3\}$, which would not be allowed when assuming CMC. So the d-connection condition—contrary to CMC—implies the independencies to be expected when applied to cyclic graphs (Pearl and Dechter 1996; Spirtes 1995). For another example of how CMC and the d-connection condition's behaviors differ from each other see Spirtes et al. (1993, p. 359).

Before we can go on, some critical remarks on CMC seem to be apropriate: Till now it is quite controversial whether CMC is (or can be) satisfied by certain systems in quantum mechanics. The problem with EPR-correlations, for example, is that the correlation between the states of two entangled particles does not vanish when conditionalizing on their common cause (cf. Hausman 1998, p. 252; Healey 2009). This would be no problem if the two states could be explained by some direct causal connection, but such a connection is excluded by the theory of relativity, or more precisely: by the fact that no causal influence can spread faster than light. Näger (2016), on the other hand, shows how CMC could be satisfied in EPR experiments in case the causal faithfulness condition (see Sect. 3.4.2) is violated. Schurz (in press) suggests a weakening of the causal Markov condition that allows for common causes that do not screen off their effects. Because there seems to be no consense about the role of CMC in the quantum world at the moment, I prefer to exclude the quantum world from the domain to which CMC should be applied.

But there are not only counterexamples of CMC-violating systems from the quantum domain, but also from the macro world. Cartwright (2007, p. 122), for example, argues that one could have a system similar to the EPR experiment in a macro domain: Assume $V = \{X, Y, Z\}$, where Z stands for a chemical reaction that produces a substance X as well as a byproduct Y. The probability of getting the intended substance ($X = 1$) given the reaction ($Z = 1$) is 0.8. However, if the reaction produces the substance, then and only then it produces the byproduct ($Y = 1$). In this scenario we have a simple common cause structure: The reaction Z is a common cause of the substance X and the byproduct Y (i.e., $X \longleftarrow Z \longrightarrow Y$). But while X and Y are correlated, this correlation does not vanish when conditionalizing on the common cause Z (because $P(X = 1|Z = 1) = 0.8$ but $P(X = 1|Y = 1, Z = 1) = 1$). Thus, CMC is violated.

Note that the EPR example as well as Cartwright's (2007) example illustrated above are typically seen as counterexamples to CMC. Strictly speaking, however, CMC is (as already mentioned earlier) formulated as a condition that may or may not hold. CMC does not say that it is satisfied by any real world system. So the scenarios described can in the best case show that assuming CMC to hold for all kinds of systems may be problematic. But even if CMC would be assumed for all kinds of systems, then these scenarios could not be straightforwardly interpreted as counterexamples to CMC. What they can show is that CMC is violated given that certain additional causal assumptions hold, such as the assumption that there are no hidden variables, the assumption that no selection bias is involved, and the assumption that there are no additional causal arrows. This means that typical counterexamples to CMC only show that there is no underlying causal structure satisfying the conjunction of CMC and these additional assumptions. In Sect. 4.3 I will briefly summarize a result from (Schurz and Gebharter 2016): Strictly speaking, there cannot be counterexamples to CMC, simply because assuming CMC to hold alone does not imply any empirical consequences. For other convincing defenses of CMC, see, e.g., (Hitchcock 2010), (Pearl 2000, p. 62), and (Spirtes et al. 2000, pp. 59–63). A novel justification of an axiom assuming CMC by an inference to the best explanation will be developed in Sect. 4.2.

3.4 Minimality and Faithfulness

3.4.1 Minimality

The *causal minimality condition* (Min) is essentially the minimality condition (Definition 2.5) adopted for causal models that satisfy the causal Markov condition (or the equivalent *d*-connection condition) (cf. Spirtes et al. 2000, p. 31):

Condition 3.3 (causal minimality condition) *A causal model $\langle V, E, P \rangle$ that satisfies the causal Markov condition (or the d-connection condition) satisfies the (causal) minimality condition if and only if there is no submodel $\langle V, E', P \rangle$ of $\langle V, E, P \rangle$ with $E' \subset E$ that satisfies CMC.*

Satisfying the minimality condition guarantees that every causal arrow in a causal model is probabilistically relevant. When deleting the arrow of a minimal causal model, the resulting causal model will not satisfy CMC anymore. Note that minimality, as it is typically introduced (see Condition 3.3), is only defined for causal models satisfying CMC. Minimality can, however, be generalized in such a way that it can be formulated without requiring CMC to hold. This generalization of the minimality condition is called the *productivity condition* (Prod) in (Schurz and Gebharter 2016, sec. 2.3):

Condition 3.4 (productivity condition) *A causal model $\langle V, E, P \rangle$ satisfies the productivity condition if and only if $DEP_P(X, Y | Par(Y) \backslash \{X\})$ holds for every $X, Y \in V$ with $X \longrightarrow Y$ in $\langle V, E \rangle$.*

The main idea behind the productivity condition is basically the same as behind the minimality condition: In a causal model $\langle V, E, P \rangle$ satisfying the productivity condition it is guaranteed that every causal arrow is probabilistically relevant, i.e., is probabilistically productive: Every $X \longrightarrow Y$ in a causal model $\langle V, E, P \rangle$ produces a probabilistic dependence between X and Y when one conditionalizes on all X-alternative parents of Y. It can be shown that productivity and minimality are equivalent for DAGs, provided CMC is satisfied (Gebharter and Schurz 2014, sec. 2.3, theorem 1).

We assert the minimality/productivity condition for all kinds of causal systems because we want our theory to make only claims which are, at least in principle, empirically testable. So we can introduce an axiom (Schurz and Gebharter 2016, p. 1087) assuming the productivity condition to hold for all kinds of causal systems and justify this axiom by a theoretical consideration, *viz.* by a version of Occam's razor: Do only postulate causal relations when they are required to explain something. Unproductive causal arrows, which do not leave any kind of empirical footprints, cannot explain anything and, thus, should be eliminated.

Axiom 3.1 (causal productivity axiom) *Every real causal system $\langle V, E, P \rangle$ satisfies the productivity condition.*

We refer to this new axiom by the label 'CPA', which stands short for 'causal productivity axiom'.

3.4.2 Faithfulness

The (causal) faithfulness condition (CFC) is crucial when it comes to causal inference on the basis of empirical data. It is typically formulated only for causal models that satisfy CMC (or the *d*-connection condition): If $\langle V, E, P \rangle$ is a causal model satisfying CMC, then $\langle V, E, P \rangle$ satisfies the (causal) faithfulness condition if and only if P does not feature a (conditional or unconditional) probabilistic independence relation not implied by $\langle V, E \rangle$ and CMC (cf. Spirtes et al. 2000, p. 31). According to this definition, faithfulness requires that the probabilistic independence relations implied by CMC are all the probabilistic independence relations featured by a causal model's probability distribution. If the causal model of interest satisfies CMC and does not feature causal cycles, faithfulness implies minimality/productivity (cf. Spirtes et al. 2000, p. 31).

The faithfulness condition can be generalized in such a way that its formulation does not require CMC to hold (cf. Schurz and Gebharter 2016, p. 1090):

Condition 3.5 (causal faithfulness condition) *A causal model $\langle V, E, P \rangle$ satisfies the causal faithfulness condition if and only if for all $X, Y \in V$ and for all $M \subseteq V \backslash \{X, Y\}$: If X and Y are d-connected given M in $\langle V, E \rangle$, then $DEP_P(X, Y | M)$.*

Condition 3.5, which is the converse of the d-connection condition (Condition 3.2), reveals the full content of the faithfulness condition (CFC): While every (conditional) probabilistic dependence in a causal system satisfying the d-connection condition can be accounted for by some d-connection in the system, every (conditional) d-connection (or causal connection) in a model satisfying the faithfulness condition produces a (conditional) probabilistic dependence.

The equivalence of Condition 3.5 and Spirtes et al.'s (2000) definition stated above when CMC (or the d-connection condition) holds is well known. Since it will often be convenient to have a notion of faithfulness that is independent of CMC (or the d-connection principle), I will rather make use of Condition 3.5 than of the original version introduced in Spirtes et al. (2000). Thus, from now on 'causal faithfulness condition' (and 'CFC') shall stand for the generalized version of the causal faithfulness condition, i.e., for Condition 3.5.

In case of acyclic causal models that satisfy CMC (or the d-connection condition), faithfulness is equivalent to *parameter stability* (cf. Pearl 2000, p. 48). For faithfulness not to hold (i.e., to be not parameter stable) a fine-tuning of the causal model of interest's parameters in such a way that probabilistic dependencies between causally connected variables disappear due to this fine-tuning is required. It is because of this that non-faithful causal systems can be expected to be extremely rare.

Definition 3.1 (parameter stability/instability) If $\langle V, E, P \rangle$ is an acyclic causal model satisfying CMC (or the d-connection condition), then $\langle V, E, P \rangle$ is parameter stable if and only if P does not feature (conditional or unconditional) probabilistic independence relations which can be destroyed by arbitrarily small changes of some of the model's parameters.

$\langle V, E, P \rangle$ is parameter instable if and only if $\langle V, E, P \rangle$ is not parameter stable.

The faithfulness condition is probably the most controversial condition of the causal nets formalism. There are several plausible possibilities for faithfulness to not hold. In the following, I will briefly discuss the most important cases of non-faithfulness.

3.4.2.1 Non-faithfulness Due to Non-minimality/Non-productivity

The first possibility for faithfulness to not hold is due to a causal system featuring non-productive arrows. From an empirical point of view it is, however, reasonable to assume only such causal relations which are (at least in principle) testable. We only assume theoretical relations when we are forced to by empirical facts. This reasoning constitutes some kind of application of Occam's razor: No entity without necessity. This theoretically justifies why we should assert that causal systems are not non-faithful due to non-minimality.

3.4.2.2 Non-faithfulness Due to Intransitivity

Non-faithfulness due to intransitivity arises if two variables X_1 and X_n are independent conditional on a subset M of the model's variable set $V\backslash\{X_1, X_n\}$, though M d-connects X_1 and X_n over a causal path $\pi : X_1 - \ldots - X_n$ such that every edge $X_i - X_{i+1}$ (with $1 \leq i < n$) that is part of this path is productive. There are basically two kinds of faithfulness violations due to intransitivity: Either there is a causal subpath $X_i - X_{i+1} - X_{i+2}$ such that X_i and X_{i+2} only depend on different X_{i+1}-values, or the parameters are fine-tuned in such a way that the probabilistic influences transported over a subpath $X_i - X_{i+1} - X_{i+2}$ cancel each other exactly. The latter kind of violation of faithfulness is called "violation of faithfulness due to internal canceling paths" by Näger (2016).

3.4.2.3 Non-faithfulness Due to Cancelation

Another possibility for faithfulness to not hold is that there may be influences on Y that exactly cancel the influence of another variable X on Y though X and Y are d-connected by some causal path π. We distinguish two kinds of this sort of non-faithfulness: non-faithfulness due to canceling paths and non-faithfulness due to canceling causes. Non-faithfulness due to canceling paths arises when the probabilistic influences transported over two or more causal paths d-connecting two variables X and Y exactly cancel each other such that $INDEP_P(X, Y)$. In case of non-faithfulness due to canceling causes there is no other path between two variables X and Y canceling the influence of some d-connection between X and Y. Here X's influence on Y over a causal path π d-connecting X and Y is canceled by the influence of another cause Z of Y. (For an example, see Pearl 1988, p. 256.)

3.4.2.4 Non-faithfulness Due to Deterministic Dependence

This kind of non-faithfulness arises in models in which some variables depend deterministically on other variables (or on sets of other variables). A variable Y deterministically depends on another variable (or a set of variables) Z if and only if for every Z-value z there is an Y-value y such that $P(y|z) = 1$ holds. Now we have a case of non-faithfulness due to deterministic dependence if (i) Y is d-connected over a path π with some variable X given Z, and (ii) Y deterministically depends on Z.

Authors such as Cartwright (1999b, p. 118) and Hoover (2001, p. 171), for example, argue that non-faithfulness due to cancelation and non-faithfulness due to intransitivity may be much more frequent than one would expect. These kinds of non-faithfulness are quite typical in self-regulatory systems such as evolutionary systems or artificial devices. Evolution has produced systems that balance external perturbations and many man-made devices are made to render a certain behavior of a system independent from external influences. An example for the latter would be an air conditioner: The outside temperature directly causes the inside temperature, which directly causes the state of an air conditioner, which, in turn, directly causes the inside temperature (i.e., outside temperature $\longrightarrow$ inside temperature $\leftrightarrows$ air conditioner). The device is built in such a way that the inside temperature should be independent from the outside temperature when the air conditioner is on. What can we say about such violations of faithfulness?

Since faithfulness is equivalent to parameter stability (recall Definition 3.1), every non-faithful (conditional) independence (of any kind) can be destroyed by slightly varying the respective model's parameters. Based on this fact, faithfulness is typically defended by the formal result that non-faithful systems have Lebesque measure zero, given the parameters of the respective system are allowed to vary (cf. Spirtes et al. 2000, pp. 41f; Steel 2006, p. 313). This result, however, can only be used as a defense of generally assuming the faithfulness condition if the parameters of all (or almost all) systems in the actual world fluctuate over time. Such parameter fluctuation can be represented by adding external noise variables to a causal model. An *external noise variable* N_X for a variable X is a variable that is exogenous, that is a direct cause of X and only of X, and that depends on some X-values x conditional on some instantiations $par(X)$ of X's parents $Par(X)$. A noise variable N_X represents all kinds of perturbations on X not represented in the model (cf. Pearl 2000, p. 27; Spirtes et al. 2000, p. 28).

Under the assumption that all noise variables are mutually independent, a causal model's parameters $P(x|par(X))$ can be changed according to the following equation by independently varying the prior distribution $P(n_X)$ over the model's noise variables:

$$P(x|par(X)) = \sum_{n_X} P(x|par(X), n_X) \cdot P(n_X) \tag{3.1}$$

We can now formulate the following *external noise assumption* postulating that almost all causal systems' parameters vary over time due to the influence of external noise (cf. Schurz and Gebharter 2016, p. 1093):

External noise assumption: The variables of (almost all) causal structures in our world are causally influenced by many small and mutually independent disturbances (external noise) that fluctuate randomly over time.

If the noise assumption is true, then non-faithfulness due to cancelation and intransitivity non-faithfulness due to internal canceling paths should almost never occur. Even the prior distributions over the noise variables of evolutionary systems and artificial devices can be expected to fluctuate (at least slightly). Already very small perturbations suffice to destroy non-faithful independencies.

Unfortunately there are still problems with non-faithfulness due to deterministic dependencies. There are two possible cases: Either our world is deterministic or it is indeterministic. In both cases non-faithfulness due to deterministic dependence may arise. If the world is indeterministic, however, we will only have "functional" deterministic dependence. This kind of deterministic dependence will vanish when the prior distribution over the involved variables' noise variables changes. Thus, the external noise assumption renders non-faithfulness due to deterministic dependence extremely improbable if the world is indeterministic. If the world is deterministic, on the other hand, this is not the case. In a deterministic world there would be no external noise variables. When conditionalizing on all causally relevant factors of a variable X, X will be determined to take a certain value x with probability 1. The problem here is that we do not know whether the world is deterministic; we have no clue about how likely it is that the external noise assumption renders non-faithfulness due to deterministic dependence highly improbable.

Another problem may arise for certain systems featuring intransitivity non-faithful independencies. It may, for example, be that the intersection of set Z_X of Z-values on which X depends and the set Z_Y of Z-values on which Y depends in a causal chain $X \longrightarrow Z \longrightarrow Y$ is empty (i.e., $Z_X \cap Z_Y = \emptyset$) due to the nature of the system's underlying causal mechanism. So we would have $DEP_P(X, Z_X)$, $DEP_P(Z_Y, Y)$, but $INDEP_P(X, Y)$ simply because $Z_X \cap Z_Y = \emptyset$.

Summarizing, it seems that the external noise assumption only renders non-faithfulness due to cancelation and intransitivity non-faithfulness due to internal canceling paths highly improbable. But non-faithfulness due to deterministic dependencies and certain versions of non-faithfulness due to intransitivity still pose a threat. In Schurz and Gebharter (2016, sec. 3.2) we argued that one can deal with these cases of non-faithfulness by a weakening of CFC. The problematic cases of non-faithfulness can be excluded by means of purely empirical conditions. Hence, the faithfulness condition should be assumed only for systems satisfying these conditions. So, instead of an axiom assuming the full faithfulness condition for all kinds of causal systems, we formulate the following *restricted causal faithfulness axiom* (RCFA):

Axiom 3.2 (restricted causal faithfulness axiom) *If $\langle V, E, P \rangle$ satisfies CMC (or the d-connection condition) and the productivity condition and $\langle V, P \rangle$ satisfies conditions (a) and (b) below, then $\langle V, E, P \rangle$ is faithful with very high probability.*

(a) No value of any $X \in V$ depends deterministically on a subset $M \subseteq V \backslash \{X\}$.
(b) There exists no sequence of variables $\langle X_1, \ldots, X_n \rangle$ such that

    (b.1) $DEP_P(X_i, X_{i+1})$ (with $1 \leq i < n$), but
    (b.2) $INDEP_P(X_1, X_n | M)$, where $M = \{X_i : INDEP_P(X_{i-1}, X_{i+1})$, where $1 < i < n\}$.

Condition (a) in Axiom 3.2 excludes non-faithfulness due to deterministic dependence and condition (b) excludes non-faithfulness due to intransitivity with high probability. Whether both conditions hold can be empirically tested. Axiom 3.2 claims that almost all causal systems not non-faithful due to deterministic dependence and not non-faithfulness due to intransitivity are faithful. This claim is justified by the external noise assumption.

Chapter 4
Causality as a Theoretical Concept

Abstract In the first part of this chapter I finish the axiomatization of the causal nets framework started in Chap. 3. I also argue that the causal Markov axiom provides the best explanation for two statistical phenomena. In the second part I present several results about the empirical content of different versions (i.e., combination of axioms) of the theory of causal nets. Both parts together show that causation satisfies the same modern standards as theoretical concepts of good empirical theories do. This can be seen as new empirical support for the theory of causal nets, but also as an answer to Hume's skeptical challenge: Actually, it seems that we have good reasons to believe in causation as something ontologically real out there in the world.

4.1 Introduction

One of the most important recipes for the evolutionary success of our species seems to be human's ability to do causal reasoning (Tomasello 2009). This ability is closely connected to our ability of influencing and manipulating our environment in compliance with our needs and desires: We have indirect control over the occurrence of certain events by directly bringing about other events. The latter we tend to regard as causes of the former, and the former as these causes' effects. The more powerful a species' causal reasoning potential becomes, the more insight it can be expected to achieve in the causal structure of the world, and hence, the better it can be expected to be able to control its environment.

Causation is not only closely tied to the possibility of bringing about events by directly manipulating other events, but also to our concept of explanation. Though not all explanations are causal explanations (cf. Friedman 1974; Kitcher 1989; Woodward 2011b), many explanations in our everyday life as well as in the sciences obviously are causal. When we want to know, for example, why there is a storm, we want to know what caused the storm. We are not satisfied with answers providing only non-causal information, not even when these answers refer to events highly correlated with the rising of a storm, such as the falling of the barometer reading, for example. The falling of the barometer reading cannot explain the storm. The barometer reading as well as the rising of the storm are correlated because there is a common cause, *viz.* the falling of the atmospheric pressure, which (causally) explains both events.

© Springer International Publishing AG 2017

A. Gebharter, *Causal Nets, Interventionism, and Mechanisms*,
Synthese Library 381, DOI 10.1007/978-3-319-49908-6_4

Now though our ability to distinguish between causes and effects makes us evolutionary successful and seems to be required for generating (causal) explanations, the concept of causation we seem to possess makes troubles since ancient times. Again and again philosophers tried to explicate the concept of causation we use so successfully, but till now it seems that almost all of their attempts were doomed to failure. Hume (1738/1975, 1748/1999) even went as far as to claim that there is no such thing as causation at all in the objective world (see also Norton 2009). All there is are regularities; the events (or types of events) involved in such regularities can be structured by us and other cognitive beings into causes and effects.[1] The better an individual can do this structuring, the better it can survive in this world. This would amount to a subjective understanding of causation.

Such a subjective understanding of causation seems to lead to problems. On the one hand it cannot explain why some events X correlated with an event Y can be used to influence Y, while others cannot. (In the barometer scenario, for example, atmospheric pressure as well as barometer reading are correlated with the coming of a storm, but a storm could only be brought about by changing the atmospheric pressure, and not by manipulating the barometer reading.) Mere regularity (understood as correlation) is symmetric. But causation (typically) is not. One can bring about the effect by manipulating the cause, but not the other way round. On the other hand, we would still not want to accept merely correlated events as explanatory. Recall, for example, the barometer scenario discussed above once again. There is a nice correlation between the falling of the barometer reading and the rising of a storm, but we do not want to accept the former as explanatory relevant for the latter. What we want is a *causal* explanation.

But even if we tend to accept these objections to Hume's claim that causation is not real, we still are in debt of providing a successful explication of causation. Most attempts to explicate the concept of causation till now were based on the idea of what Glymour (2004) calls a *Socratic account*. Socratic accounts try to explicate the meaning of a concept by providing necessary conditions which, if taken together, are also sufficient for something to fall under that concept. Some of the more prominent reductive Socratic attempts to explicate the meaning of causation are, for example, Collingwood (2002), Gasking (1955), Good (1959), Graßhoff and May (2001), Lewis (1973), Mackie (1965), Mackie (1974), Menzies and Price (1993), Psillos (2009), Reichenbach (1956/1991), Suppes (1970), Salmon (1997), Dowe (2007), von Wright (1971).

Others, after recognizing that all reductive attempts to explicitly define causation failed so far, try to characterize causation by means of other causal notions, but still in the form of necessary and sufficient conditions. Some of the more prominent accounts are, for instance, Cartwright (1979, 1989), Eells (1987), Eells and Sober (1983), Glennan (1996), Hausman (1998), Skyrms (1980), Woodward (2003),

---

[1]Modern philosophers who see causation as subjective are, for example, Spohn (2001, 2006) and Williamson (2005). Spohn explicitly presents causation in a Kantian manner as a projection. I am indebted to Markus Schrenk for pointing this out to me.

Woodward and Hitchcock (2003), Hitchcock and Woodward (2003). No one of these accounts is commonly accepted until now, and this for good reasons (see, for example, the discussion of the diverse approaches and their problems in Beebee et al. 2009). On the other hand, there are also *Euclidic accounts* (cf. Glymour 2004). These accounts content themselves with less than providing necessary and sufficient conditions. They explicate the meaning of a concept by defining it implicitly; they connect this concept by several axioms to other concepts.

If we take a look at how theoretical concepts (cf. Balzer et al. 1987; Carnap 1956; French 2008; Hempel 1958; Lewis 1970; Papineau 1996; Sneed 1979) are treated in the sciences, we find that these concepts are introduced nearly exclusively in a Euclidean fashion. The concept of force in Newtonian physics, for example, is not defined, but characterized by several axioms which connect it to empirically easier accessible concepts. Now the idea is that maybe also metaphysical concepts (such as causation) are best analyzed in this vein. The most promising Euclidean approach to causation till now seems to be the causal Bayes net framework (Pearl 2000; Spirtes et al. 2000). The approach was mainly developed with the methodological aim of discovering causal structures on the basis of empirical data in mind. It provides several conditions that connect causal relations to probability distributions (whenever these conditions are satisfied). There is also a lot of work done on how the framework can be used for causal reasoning and how its models can be used to predict the effects of human (or non-human) interventions. But till now the approach's philosophical status is not quite clear. Some seem to see it as a real theory of causation and not just as a methodological tool (cf. Spirtes et al. 2000, sec. 1). Others (most prominently Cartwright 2001, 2007) argue against such a view or at least consider the theory as inadequate as a general theory of causation.

In (Schurz and Gebharter 2016) we suggested an axiomatization of the causal Bayes net framework and investigated whether causation, as it is characterized within the resulting theory, satisfies accepted standards for theoretical concepts. Recall the analogy with force in Newtonian mechanics mentioned above. We belief that forces are real and that Newtonian mechanics correctly characterizes force because the concept of force satisfies such standards. It provides the best available explanation for a multitude of phenomena, it unifies phenomena of different domains, it allows to generate predictions (and also novel predictions), and Newtonian mechanics has empirical consequences by whose means it can be tested as a whole theory (at least when the law of gravitation is added). Having all of these characteristics is typically seen as lending strong support to the belief that the respective concept corresponds to something ontologically real. In (Schurz and Gebharter 2016) we showed that causation, as it is characterized within the theory of causal nets, satisfies the same standards. This somehow answers Hume's challenge. It lends strong support to causation as something out there in the world that is, though not explicitly defined, correctly described by the theory of causal nets. Maybe this is as much as we can hope to achieve to know about causation.

In this section I summarize the results obtained in Schurz and Gebharter (2016). I also present some new results and explain some of the results from the mentioned paper in much more detail. Especially the step by step inference to the best explanation to arrive at an axiom presupposing the causal Markov condition to hold is much more detailed in this book. In the mentioned paper, we have only sketched how the inference would work in case of simple causal systems consisting of three variables and without any latent common causes. In Sect. 4.2 I present results concerning the theory of causal nets' explanatory warrant, and in Sect. 4.3 results about its empirical content. The basic idea underlying Sect. 4.2 is to use Descartes' method and try to put away all our causal intuitions as well as all our knowledge about causal theories and concepts we have acquired in our life and, instead, focus only on observations and empirical data. I then investigate the question of whether there are empirical facts which force us to introduce causal concepts to explain these facts. In particular, I will take a closer look at certain empirical systems which actually can be found in the world. We can define such an empirical system as follows:

Definition 4.1 (empirical system) $\langle V_e, P_e \rangle$ is an empirical system if and only if V_e is a set of analytically independent[2] empirically measurable variables and P_e is a probability distribution over V_e.

Most of the time we will only consider robust probability distributions. That a probability distribution P over a variable set V is *robust* means that no probabilistic independence relation implied by this probability distribution P disappears when slightly varying the conditional probabilities of the variables in V by means of external perturbations. Such not disappearing independence relations are called robust; disappearing independence relations, on the other hand, are called non-robust. Essentially, assuming robustness of a probability distribution amounts to assuming parameter stability (or faithfulness; see also Sect. 3.4.2).

It will turn out that some empirical systems possess certain quite surprising opposite statistical properties, *viz.* screening off and linking-up, whose occurrence begs for an explanation. Presupposed that these statistical properties are not some kind of miracle, there is only one good explanation for these phenomena available: the assumption of certain relations (we are forced to characterize in a very special way) which produce these properties. So we are forced to introduce a certain concept (which we could call "causation") to explain otherwise unexplainable phenomena. We proceed by an inference to the best explanation (see Schurz 2008 for a general approach) to justify causality as an objective feature of the world and, thereby, arrive at a certain theory of causation whose core axiom assumes the causal Markov condition (recall Sect. 3.3) to hold. So the justification of causality presented in this section can also be seen as independent support for an axiom asserting that the causal Markov condition holds in certain systems.

The structure of Sect. 4.2 is as follows: In Sect. 4.2.1 I ask whether pure correlations are evidence enough for assuming causal relations. In Sect. 4.2.2 I present the first one of the above-mentioned statistical properties: screening off.

---

[2]This constraint is required to distinguish synthetic causal dependencies from analytic dependencies such as dependencies due to meaning, definition, conceptualization, etc.

It will turn out that the explanation of screening off phenomena forces us to assume causal structures which have to meet certain conditions. In Sect. 4.2.3 it will turn out that our characterization of causation, as developed in Sect. 4.2.2, needs further refinement to also explain another statistical property: linking-up. In Sect. 4.2.4 I will develop a (preliminary) axiom characterizing causal relations in such a way that they can also explain more complex scenarios in which both, screening off as well as linking-up, are present. In Sect. 4.2.5 I investigate how a slightly weaker version of this preliminary axiom can be used to explain screening off and linking-up without robust probability distributions. This axiom—to which I will refer as the causal Markov axiom (CMA)—will turn out to presuppose the causal Markov condition (CMC) to hold.

In the second part of Chap. 4, i.e., Sect. 4.3, I present results concerning the resulting theory of causal nets' empirical content. I start by presenting results about the empirical content of CMA (alone) in Sect. 4.3.1. In Sect. 4.3.2 I ask the same question for a theory version asserting the causal faithfulness condition (CFC). It will turn out that neither CMA alone nor assuming CFC alone implies any empirical consequences. However, combining CMA and generally assuming CFC leads to deductive empirical content, while combining CMA and the restricted causal faithfulness axiom (RCFA) leads to probabilistic empirical content (Sect. 4.3.3). One gets strong empirical content by adding the assumption that causes always precede their effects in time (Sect. 4.3.4). In Sect. 4.3.5 I finally investigate the question what the assumption that (most) human actions are free would contribute to the theory's empirical consequences.

4.2 Explanatory Warrant of the Theory of Causal Nets

4.2.1 Explaining Correlation

Let us begin our venture by asking whether pure correlations implied by an empirical model $\langle V_e, P_e \rangle$'s probability distribution P_e force us to postulate causal relations with certain properties to explain them. We start with the most simple case: empirical systems whose variable sets contain only two variables X and Y. When we look at several empirical systems of this kind, then we will find that X and Y are probabilistically dependent in some but not in all empirical systems' probability distributions. Smoking and lung cancer, for example, will be correlated, while birth rate and couch drop consumption will (presumably) be probabilistically independent. This observation requires an explanation: Why are some empirically measurable variables X and Y of some empirical systems correlated, while others of other empirical systems are not? One may answer this question by postulating a causal connection between those variables X and Y which are probabilistically dependent, while such a causal connection does not exist between those which are uncorrelated. So the idea here would be that every correlation between two variables

X and *Y* is produced by a causal connection between these variables, and that *X* and *Y* are probabilistically independent if no such productive relation exists. So one would get the following (preliminary) axiom connecting causation to observable phenomena:

Preliminary Axiom 4.1 *If $\langle V_e, P_e \rangle$ is an empirical system, then for all $X, Y \in V_e$: $DEP_{P_e}(X, Y)$ if and only if X and Y are causally connected in V_e, i.e., X and Y are connected by a binary causal relation.*

Causation, as introduced by this (preliminary) axiom (alone), does clearly not fulfill the discussed requirements for reasonable theoretical concepts. Preliminary Axiom 4.1 just postulates a (not further specified) causal connection for every probabilistic dependence and only for probabilistic dependencies between two variables *X* and *Y*, and thus, just postulates a causal connection for every measured correlation which can only ex post explain such correlations. We have just an empirically empty duplication of entities: For every probabilistic dependence relation between two variables there is a corresponding causal relation and vice versa. If we represent correlations by dotted lines ('· · ·') and causal connections by continuous lines ('—'), then we can illustrate this finding on the three exemplary empirical systems in Fig. 4.1.

So far, our observations do not force us to make any specifications of causal connection going beyond the weak condition that causal connection is symmetric (this seems to be a direct consequence of Preliminary Axiom 4.1 and the fact that probabilistic dependence is symmetric). In Sect. 4.2.2 we will see that there are more complex probabilistic properties of empirical systems which actually force us to add further refinements to our notion of causal connection under development in such a way that it becomes empirically significant.

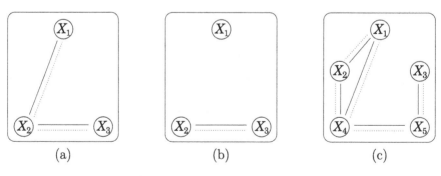

                    (a)   (b)   (c)

Fig. 4.1 Preliminary Axiom 4.1 (without any further specification of causal connection) leads to a duplication of correlation relations. '—' stands for causal connection and '· · ·' for correlation

4.2.2 Explaining Screening Off

Sometimes we find that two formerly correlated variables become probabilistically independent after holding a third variable (or a set of variables) fixed at an arbitrarily chosen value. This probabilistic *screening off* property can be defined as follows (Schurz and Gebharter 2016, p. 1078):

Definition 4.2 (screening off) If P is a probability distribution over variable set V, then for all $X, Y \in V$ and $M \subseteq V \backslash \{X, Y\}$: X is screened off from Y by M if and only if

(a) $DEP_P(X, Y)$, and
(b) $INDEP_P(X, Y|M)$.

A typical screening off scenario is the famous barometer scenario (cf. Grünbaum 1962). The falling of the barometer reading $B = b$ and the rising of a storm $S = s$ are positively correlated, but this correlation breaks down after the actual value p of the atmospheric pressure P is known (i.e., is held fixed). We are, even if not knowingly, so familiar with the probabilistic consequences of causal structures that we unhesitatingly tend to explain this disappearance of the correlation between B and S with a common cause structure: B and S are uncorrelated after conditionalizing on P because P is the common cause of B and S—so knowing that the air pressure is p renders any additional information concerning the barometer reading's actual value b probabilistically irrelevant for the rising of a storm $S = s$.

However, just for the moment, try to let go of this strong causal intuition and assume that we do only observe that the barometer scenario possesses the screening off property as defined above. Then we can ask ourselves why this property appears in the barometer scenario, but not in other scenarios. There is, for instance, no screening off in a scenario consisting of the three variables birthrate, amount of storks, and couch drop consumption. While birthrate and amount of storks are correlated (cf. Schurz 2013, p. 206), this correlation will (presumably) not break down after conditionalizing on any possible value of the variable couch drop consumption. But why do we have screening off in the barometer scenario?

The best explanation of the screening off phenomenon seems to be that probabilistic dependencies are produced by causal pathways consisting of one or more binary causal relations (cf. Schurz and Gebharter 2016, sec. 2.2). In the case of the barometer scenario, B and P as well as P and S are connected by binary causal relations (which shall be represented by continuous lines '—'), but not B and S. So the probabilistic influence of B on S is transported over the causal path $B - P - S$ (see Fig. 4.2 for a graphical illustration). In other words this means that B-variations can lead to S-variations only over P-variations. Thus, if P is fixed to a certain value p, then S is probabilistically insensitive to any possible B-variation, and thus, B is screened off from S by P. This explanation of screening off sheds some light on how 'causal connection' in Preliminary Axiom 4.1 could be further specified; a causal connection is required to be a chain of binary causal relations:

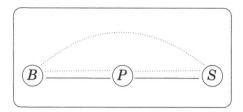

Fig. 4.2 Barometer reading B and rising of a storm S are connected via a path of binary causal relations ('—') going through atmospheric pressure P. Each causal relation as well as every causal path ('— ... —') produces a probabilistic dependence ('$\cdots$') between the variables it connects. So not only B and P as well as P and S, but also B and S are correlated ($B \cdots S$). But since this probabilistic dependence is only due to the causal connection $B - P - S$, B-variations can lead to S-variations (and vice versa) only over P-variations. Hence, the probabilistic dependence between B and S breaks down when conditionalizing on P

Preliminary Axiom 4.2 *If $\langle V_e, P_e \rangle$ is an empirical system, then for all $X, Y \in V_e$: $DEP_{P_e}(X, Y)$ if and only if X and Y are causally connected in V_e, i.e., connected via a path $\pi : X - \ldots - Y$ of binary causal relations in V_e.*

Note that till now we are (by inference to the best explanation) only forced to introduce binary causal relations; there is no need to assume that causal relations are directed, i.e., there is no need to distinguish between cause and effect so far. Do also note that the if- as well as the only-if-direction in Preliminary Axiom 4.2 is required: The if-direction is needed to assure that no correlation occurs without causal connection, and the only-if-direction is required to guarantee that causal structures of the form $X - Z - Y$ do produce screening off, for which, according to condition (a) of Definition 4.2, an X-Y correlation is required. Preliminary Axiom 4.2 does, however, not yet explain why X and Y become probabilistically independent when conditionalizing on Z in causal structure $X - Z - Y$; this is, according to condition (b) of Definition 4.2, also required for screening off. This requirement can be captured by the following advancement of Preliminary Axiom 4.2:

Preliminary Axiom 4.3 *If $\langle V_e, P_e \rangle$ is an empirical system, then for all $X, Y \in V_e$ and $M \subseteq V_e \backslash \{X, Y\}$: $DEP_{P_e}(X, Y|M)$ if and only if X and Y are causally connected given M in V_e, i.e., connected via a path $\pi : X - \ldots - Y$ of binary causal relations in V_e such that π does not go through M.*

The main difference between Preliminary Axiom 4.3 and Preliminary Axiom 4.2 is that the former uses the notions of probabilistic dependence and causal connection in a broader sense, *viz.* relative to a set of variables M. Recall that $DEP_P(X, Y|\emptyset)$ is equivalent to $DEP_P(X, Y)$ (see Sect. 2.5). Hence, Preliminary Axiom 4.2 is a special case of Preliminary Axiom 4.3: When M is the empty set, then two variables X and Y are correlated if and only if they are causally connected via a (i.e., at least one) path π of binary causal relations (not going through $M = \emptyset$). Since every such causal

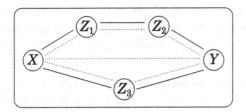

Fig. 4.3 X and Y are connected via two different causal paths, *viz.* via $X — Z_1 — Z_2 — Y$ and via $X — Z_3 — Y$, which both produce a correlation between X and Y ($X \cdots Y$). But X-variations can probabilistically influence Y only via Z_1- and Z_2- or via Z_3-variations. Thus, X and Y are screened off by $\{Z_1, Z_2, Z_3\}$, $\{Z_1, Z_3\}$, and $\{Z_2, Z_3\}$. $\{Z_1\}$, $\{Z_2\}$, and $\{Z_1, Z_2\}$ alone do not screen X and Y off, because X-variations will still influence Y due to Z_3-variations. $\{Z_3\}$ does not screen X and Y off, because X-variations can still influence Y due to Z_1- and Z_2-variations

path produces probabilistic dependence, this correlation between X and Y will only vanish if at least one variable of every causal path connecting X and Y is fixed to a certain value (see Fig. 4.3 for a graphical illustration).

Preliminary Axiom 4.3 determines when two variables X and Y of an empirical system $\langle V_e, P_e \rangle$'s variable set V_e are connected by a binary causal relation:

Theorem 4.1 (direct causal connection) *If $\langle V_e, P_e \rangle$ is an empirical system, then for all $X, Y \in V_e$: $X — Y$ in V_e if and only if $DEP_{P_e}(X, Y|M)$ holds for all $M \subseteq V_e \backslash \{X, Y\}$.*

Proof Assume $\langle V_e, P_e \rangle$ is an empirical system and $X, Y \in V_e$.

The if-direction: Assume $DEP_{P_e}(X, Y|M)$ holds for all $M \subseteq V_e \backslash \{X, Y\}$. It follows that $DEP_{P_e}(X, Y|M')$ does hold for $M' = V_e \backslash \{X, Y\}$. Thus, with Preliminary Axiom 4.3, it follows that X and Y are connected by a causal path not going through M'. The only possible such causal path in V_e is $X — Y$.

The only-if-direction: Assume $X — Y$ in V_e. Then X and Y are connected by a causal path not going through any set $M \subseteq V_e \backslash \{X, Y\}$. Thus, according to Preliminary Axiom 4.3, $DEP_{P_e}(X, Y|M)$ holds for every $M \subseteq V_e \backslash \{X, Y\}$.   □

With the help of this theorem one can directly read an empirical system's underlying causal structure off from this empirical system's probability distribution: Draw a causal connection ('—') between two variables X and Y if and only if X and Y are correlated in P_e and no subset M of the empirical system's variable set V_e not containing X and Y screens X off from Y. Here is an example: Assume we have measured the probability distribution P_e with the following probabilistic dependence and independence features over a variable set $V_e = \{X, Y, Z_1, Z_2, Z_3\}$: X and Z_1, X and Z_3, Z_2 and Z_3, Y and Z_2, and Y and Z_3 are correlated and probabilistically dependent conditional on every subset M of V_e not containing the respective pair of variables. One gets the system's underlying causal structure by drawing a direct causal connection '—' between every one of these pairs of variables. The resulting causal structure is the one depicted in Fig. 4.3.

Note that Preliminary Axiom 4.3 already does have empirical content. Preliminary Axiom 4.3 implies that probabilistic dependence produced by direct causal relations is transitive, and thus, excludes certain logically possible probability distributions P_e:

Theorem 4.2 (empirical content) *Preliminary Axiom 4.3 excludes an empirical system* $\langle V_e, P_e \rangle$ *if* V_e *contains* $X, Z_1, \ldots, Z_n, Y$ *such that* $DEP_{P_e}(X, Z_1|M)$ *holds for all* $M \subseteq V_e \backslash \{X, Z_1\}$, $DEP_{P_e}(Z_i, Z_{i+1}|M)$ *holds for all* $M \subseteq V_e \backslash \{Z_i, Z_{i+1}\}$ *(where* $i \in \{1, \ldots, n-1\}$), $DEP_{P_e}(Z_n, Y|M)$ *holds for all* $M \subseteq V_e \backslash \{Z_n, Y\}$, *and* $INDEP_{P_e}(X, Y)$.

Proof Assume $\langle V_e, P_e \rangle$ is an empirical system containing $X, Z_1, \ldots, Z_n, Y$ such that $DEP_{P_e}(X, Z_1|M)$ holds for all $M \subseteq V_e \backslash \{X, Z_1\}$, $DEP_{P_e}(Z_i, Z_{i+1}|M)$ holds for all $M \subseteq V_e \backslash \{Z_i, Z_{i+1}\}$ (where $i \in \{1, \ldots, n-1\}$), $DEP_{P_e}(Z_n, Y|M)$ holds for all $M \subseteq V_e \backslash \{Z_n, Y\}$, and $INDEP_{P_e}(X, Y)$. Then, according to Theorem 4.1, $X - Z_1 - \ldots - Z_n - Y$ is part of $\langle V_e, P_e \rangle$'s underlying causal structure. Thus, X and Y are connected by a path π of binary causal relations in V_e such that no variable on this path is an element of $\emptyset$. Therefore, according to Preliminary Axiom 4.3, $DEP_{P_e}(X, Y|\emptyset)$ holds, what contradicts $INDEP_{P_e}(X, Y)$.   □

So a theory of causation that characterizes causation by means of Preliminary Axiom 4.3 is empirically testable as a whole theory. If one of the excluded but logically possible empirical systems described in Theorem 4.2 would be found in the world, then this theory would be falsified. If, on the other hand, the number of observed empirical systems not having the excluded properties would increase, then this theory would become better and better confirmed.

4.2.3 Explaining Linking-Up

In the last subsection we saw that the observation that some empirical systems possess the screening off property (recall the barometer scenario discussed in Sect. 4.2.2) led us, by an inference to the best explanation, to the assumption that there are certain causal structures consisting of binary causal relations which produce mere correlation as well as screening off. We were forced to specify 'causal connection' as in Preliminary Axiom 4.3 and saw that this characterization leads to a theory of causation which already has some empirical consequences (see Theorem 4.2). Let us now take a brief look at these empirical consequences and check whether some empirical systems which are excluded by Preliminary Axiom 4.3 can be found in the world—so let us test the theory of causation we have developed so far.

According to Theorem 4.2, Preliminary Axiom 4.3 excludes (among others) empirical systems $\langle V_e, P_e \rangle$ containing three variables X, Y, and Z such that X and Z as well as Y and Z are correlated conditional on every subset of V_e not containing these variables, while X and Y are unconditionally probabilistically independent ($INDEP_{P_e}(X, Y)$). Unfortunately, there is an empirical system $\langle V_e, P_e \rangle$ which actually possesses these specific probabilistic properties excluded by Preliminary Axiom 4.3, *viz.* the flagpole scenario oftentimes discussed in the philosophy of science literature (see, e.g., Schurz 2001). This system's variable set V_e consists of three variables: A for the solar altitude, H for the hight of a flagpole, and L for the length of the flagpole's shadow. Since the solar altitude as well as the hight of the flagpole are causally relevant for the length of this flagpole's shadow, and the solar altitude as well as the hight of the flagpole can vary independently, we will find that A and L are correlated unconditionally (lower solar altitude will, for example, be correlated with increased shadow length; $DEP_{P_e}(A, L)$) as well as conditional on H ($DEP_{P_e}(A, L|H)$), that H and L are probabilistically dependent unconditionally (decreased flagpole hight will, for instance, be correlated with decreased shadow length; $DEP_{P_e}(H, L)$) as well as relative to A ($DEP_{P_e}(H, L|A)$), and that A and H are probabilistically independent unconditionally ($INDEP_{P_e}(A, H)$) in this scenario. Since we can compute the solar altitude a from the hight h of the flagpole as well as the hight of the flagpole h from the solar altitude a when we know the flagpole's shadow's length l, A and H are correlated conditional on L ($DEP_{P_e}(A, H|L)$).[3]

We now know some of $\langle V_e, P_e \rangle$'s probabilistic dependencies and independencies due to causally laden considerations. Let us now, according to Descartes' method, try to forget all the causal information we used to achieve this knowledge about P_e and see what the notion of causation we developed so far on purely observational grounds implies for the flagpole scenario $\langle V_e, P_e \rangle$. Since $DEP_{P_e}(A, L)$ and $DEP_{P_e}(A, L|H)$, Theorem 4.1 implies $A - L$, and since $DEP_{P_e}(H, L)$ and $DEP_{P_e}(H, L|A)$, Theorem 4.1 implies $H - L$. Theorem 4.1 does, because of $INDEP_{P_e}(A, H)$, also imply that there is no direct causal connection between A and H. So $\langle V_e, P_e \rangle$'s underlying causal structure is $A - S - H$. But, according to Preliminary Axiom 4.3, $DEP_{P_e}(A, H)$ holds because A and H are connected by the causal path $A - S - H$. This contradicts $INDEP_{P_e}(A, H)$. So the discovery of the flagpole scenario's probabilistic properties (as described above) falsifies Preliminary Axiom 4.3.

The flagpole system $\langle V_e, P_e \rangle$ possesses a statistical feature which is quite the opposite of screening off. While two correlated variables X and Y become probabilistically independent after conditionalizing on a third variable (or a set of variables) Z in screening off scenarios, two formerly uncorrelated variables X and Y become probabilistically dependent when conditionalizing on a third variable (or a set of variables) Z in the flagpole scenario. This property is called *linking-up* in Schurz and Gebharter (2016, p. 1080). It can be defined as follows:

Definition 4.3 (linking-up) If P is a probability distribution over variable set V, then for all $X, Y \in V$ and $M \subseteq V \backslash \{X, Y\}$: X and Y are linked up by M if and only if

(a) $INDEP_P(X, Y)$, and
(b) $DEP_P(X, Y|M)$.

---

[3]For a more detailed explanation why we would find exactly the mentioned probabilistic dependence and independence relations, see Gebharter (2013, p. 66).

We were forced to characterize causation by means of Preliminary Axiom 4.3 to explain screening off, but now we see that there are some real empirical systems which do possess the quite opposite linking-up property which cannot be explained by Preliminary Axiom 4.3, and, much worse, conflict with Preliminary Axiom 4.3. So how should we proceed? We could take the progress of scientific theories as a role model and not completely reject our so far developed characterization of causation, but rather add some refinements in such a way that it can, while still explaining screening off, also account for linking-up.

Let us reconsider the problem we have to solve here: There are empirical systems $\langle V_e, P_e \rangle$ with $V_e = \{X, Y, Z\}$ such that X and Y are screened off by Z in some of these systems, but are linked up by Z in others. Both statistical properties presuppose, according to Preliminary Axiom 4.3, the same causal structure $X - Z - Y$. In screening off scenarios, this causal structure produces $DEP_{P_e}(X, Y)$ and $INDEP_{P_e}(X, Y|Z)$, in linking-up scenarios, on the other hand, it produces $INDEP_{P_e}(X, Y)$ and $DEP_{P_e}(X, Y|Z)$. This observation requires an explanation. Until now we have not distinguished between cause and effect. In Sect. 4.2.2 there was simply no empirical reason for such a distinction. But maybe the discovery of this subsection that some real systems possess the linking-up property provides such a reason and the distinction between cause and effect does help us to distinguish screening off from linking-up properties on the basis of the diverse empirical systems' underlying causal structures.

From now on, one-headed arrows ('$\longrightarrow$') shall replace the continuous lines ('$-$') we used for representing direct causal connections. In addition to our notion of an empirical system $\langle V_e, P_e \rangle$, we will also use the notion of a causal graph (see Sect. 3.2) from now on.

Let us come back to our initial question of how we can explain screening off and linking-up on the basis of the distinction between cause and effect now. We already saw that an empirical system $\langle V_e, P_e \rangle$ with $V = \{X, Y, Z\}$ has the underlying causal structure $X - Z - Y$ if Z screens X off from Y or if Z links X and Y up. So there are actually four possible causal structures $\langle V_e, E \rangle$ which are candidates for underlying such empirical systems $\langle V_e, P_e \rangle$:

(a) $X \longrightarrow Z \longrightarrow Y$
(b) $X \longleftarrow Z \longleftarrow Y$
(c) $X \longleftarrow Z \longrightarrow Y$
(d) $X \longrightarrow Z \longleftarrow Y$

The challenge is to determine which of these causal structures produce screening off and which produce linking-up. For the following considerations I assume that Y is monotonically positively dependent on X whenever $X \longrightarrow Y$. This assumption is made for the sake of simplicity; without it, the following argumentation would be much more complex, but not different in principle. I also presuppose that (a)–(d) are the only causal paths connecting X and Y. I will consider more complex causal connections among X and Y later on in Sect. 4.2.4.

Let us begin with possible causal structure (a) now: In a domain D of individuals u high X-values will produce high Z-values which will produce high Y-values, and thus, X and Y will be probabilistically dependent $(DEP_{P_e}(X, Y))$ in an empirical system $\langle V_e, P_e \rangle$ with underlying causal structure $X \longrightarrow Z \longrightarrow Y$. Now suppose we conditionalize on any arbitrarily chosen Z-value z, i.e., we look at a subpopulation D_z of D, where D_z contains all the individuals $u \in D$ with Z-value z. Now probabilistic dependence of X on Y (and vice versa) requires that Z's value is allowed to vary. But in subpopulation D_z of D, Z's value is fixed (to z) and no variation of X-values will lead to a variation of Y-values (and vice versa). So we will find that X and Y are probabilistically independent whenever Z's value is fixed $(INDEP_{P_e}(X, Y|Z))$. Therefore, $X \longrightarrow Z \longrightarrow Y$ produces screening off: It produces $DEP_{P_e}(X, Y)$ and $INDEP_{P_e}(X, Y|Z)$ (see Definition 4.2). The same consideration can be applied to possible causal structure (b); one just has to swap X and Y in that case. Thus, also $X \longleftarrow Z \longleftarrow Y$ produces screening off.

What if our empirical system $\langle V_e, P_e \rangle$'s underlying causal structure is (c), i.e., $X \longleftarrow Z \longrightarrow Y$? In that case high Z-values will lead to high X- as well as to high Y-values in our population D. So if we observe high X-values, these high X-values were probably produced by high Z-values which also will have produced high Y-values. Hence, individuals u will have higher Y-values in subpopulations D_x of D with high X-values, and thus, X and Y will be correlated in P_e $(DEP_{P_e}(X, Y))$. Now assume D_z to be the subpopulation of D whose individuals u possess a certain Z-value z. In this subpopulation D_z of D, no variation of X-values will lead to variations of Y-values (and vice versa) because all probabilistic influence between X and Y is mediated via Z. We will, again, find that X and Y are probabilistically independent if Z's value is not allowed to vary $(INDEP_{P_e}(X, Y|Z))$, and thus, also $X \longleftarrow Z \longrightarrow Y$ produces screening off.

The last possible case is causal structure (d): $X \longrightarrow Z \longleftarrow Y$. Here high X-values as well as high Y-values cause high Z-values in population D. If we take a look at a subpopulation D_x of D in which all individuals u take a certain X-value x, Y-values y will be equally distributed in this subpopulation like in D—while higher X-values produce also higher Z-values, they have no causal influence on the distribution of Y-values in D. The same holds w.r.t. X and for any subpopulation D_y of D in which all individuals u take a certain Y-value y: X-values x will be equally distributed in this subpopulation D_y as in D; while higher Y-values produce also higher Z-values, they have no influence on the distribution of X-values in D. So X and Y will be uncorrelated $(INDEP_{P_e}(X, Y))$ in case $\langle V_e, P_e \rangle$'s underlying causal structure is $X \longrightarrow Z \longleftarrow Y$. But now assume we are looking at a certain subpopulation D_z of D with fixed Z-values z. In case these Z-values z are very high, then these high Z-values can only be due to high X-values in case the Y-values are very low in D_z, and vice versa: If these Z-values z are very high, then these high Z-values can only be due to high Y-values in case X-values are very low in D_z. So X and Y will be probabilistically sensitive in subpopulation D_z, and thus, $DEP_{P_e}(X, Y|Z)$. Hence, $X \longrightarrow Z \longleftarrow Y$ produces linking-up.

Summarizing, we found that causal structures (a)–(c) produce screening off and that causal structure (d) produces linking-up in an empirical system with variable set $V_e = \{X, Y, Z\}$. But we have not thought about more complex causal structures until now, especially not about causal structures in which variables are connected by more complex causal paths including intermediate causes, common causes, and common effects as well as about variables which are connected by more than one causal path. Before we will investigate such more complex causal structures and their implications for their corresponding empirical systems' probability distributions, I want to briefly discuss another important question in this subsection: How does it come that symmetric correlations $X \cdots Y$ are produced by binary causal relations, though causal relations $X \longrightarrow Y$ are in some sense asymmetric? The cause brings about its effect, while the effect does not bring about its cause. This is a crucial assumption also used for the explanations given for screening off and linking-up in empirical systems with three variables above. Here comes the answer: Assume $\langle V_e, P_e \rangle$ is an empirical system with $V_e = \{X, Y\}$ and underlying causal structure $X \longrightarrow Y$. Since high X-values cause high Y-values, there will be more individuals u with higher Y-values in a subpopulation D_x of D with high X-values x. But why will also a subpopulation D_y, in which individuals u have high Y-values, have increased X-values, i.e., higher X-values than in D? The answer is simply that many (or at least some) individuals u in D_y will have high Y-values because these high Y-values are caused by high X-values, and thus, the frequency of individuals with high X-values will be higher in D_y than in D.

4.2.4 Explaining Complex Scenarios with Possible Latent Common Causes

We still investigate the question of how 'causal connection' in Preliminary Axiom 4.3 must be characterized such that causal structures consisting of binary causal relations explain (i) why certain variable pairs in empirical systems are correlated or uncorrelated (see Sect. 4.2.1), (ii) why some correlated pairs of variables are screened off by certain subsets of an empirical system's variable set (see Sect. 4.2.2), and (iii) why some uncorrelated pairs of variables become probabilistically dependent after conditionalizing on certain subsets of an empirical system's variable set (see Sect. 4.2.3). So we search for a completion of the following principle that can explain (i)–(iii):

(*) If $\langle V_e, P_e \rangle$ is an empirical system, then for all $X, Y \in V_e$ and $M \subseteq V_e \backslash \{X, Y\}$: $DEP_{P_e}(X, Y | M)$ if and only if X and Y are causally connected by M in V_e, i.e.,
     ...

A causal connection between two variables X and Y consisting of directed binary causal relations must have one of the following forms:

(a) $X \longrightarrow \longrightarrow Y$
(b) $X \longleftarrow \longleftarrow Y$
(c) $X \longleftarrow \longrightarrow Y$
(d) $X \longrightarrow \longleftarrow Y$

Fig. 4.4 Each causal path connecting X and Y produces probabilistic dependence between X and Y $(X \cdots Y)$. One has to conditionalize on $\{Z_1, Z_2, Z_3\}$ to screen X and Y off each other

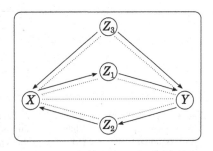

As we already saw in Sect. 4.2.3, the arrows in structures $X \longrightarrow Z \longrightarrow Y$, $X \longleftarrow Z \longleftarrow Y$, and $X \longleftarrow Z \longrightarrow Y$ probabilistically behave exactly as the asymmetric causal relations we introduced in Sect. 4.2.2; they produce probabilistic dependence between X and Y. This behavior can be generalized to directed causal chains $X \longrightarrow\longrightarrow Y$ and $X \longleftarrow\longleftarrow Y$ as well as to common cause paths $X \longleftarrow\longrightarrow Y$ of arbitrary length by means of the same argumentation as given in Sect. 4.2.3 for $X \longrightarrow Z \longrightarrow Y$, $X \longleftarrow Z \longleftarrow Y$, and $X \longleftarrow Z \longrightarrow Y$: If two variables X and Y are connected via a directed or a common cause path (and by no other causal paths), then X and Y are screened off whenever one or more of the variables lying on this path are not allowed to vary.

If X and Y are connected via more than one directed or common cause path, then the situation becomes a little bit more complicated. Since all directed and common cause paths which connect X and Y produce probabilistic dependence between X and Y, one has to fix at least one variable on each of these paths to screen X and Y off each other. This can be made clear by the following consideration: Assume the causal structure depicted in Fig. 4.4 underlies an empirical system $\langle V_e, P_e \rangle$ with $V_e = \{X, Y, Z_1, Z_2, Z_3\}$. Assume further that Z_1 is fixed to a certain value z_1 for all individuals in subpopolation D_{z_1} of D. Then, since Y is causally relevant for X over $Y \longrightarrow Z_2 \longrightarrow X$, high Y-values will still lead to higher X-values in D_{z_1}. If one holds also Z_2 fixed at a certain value z_2, i.e., if we take a look at the subpopulation $D_{z_1} \cap D_{z_2}$, higher X-values will still lead to higher Y-values because higher X-values have been produced by higher Z_3-values, which have also produced higher Y-values. So conditionalizing on Z_1 and Z_2 is insufficient for screening X and Y off from each other. But if also Z_3 is fixed to a certain value, then all causal paths between X and Y are blocked and no probabilistic influence of X can reach Y and vice versa.

But what if X and Y are connected by a directed or a common cause path *and* by a common effect path? So what if a system's underlying causal structure is, for instance, the one depicted in Fig. 4.5? Would X and Y turn out to be correlated or probabilistically independent in such a case? Here is the answer: X and Y would be correlated because in a subpopulation D_x of D whose individuals have high X-values x many (or at least some) of these individuals will have caused increased Z_1-values which will have caused increased Y-values. If one conditionalizes on Z_1, however, the probabilistic dependence between X and Y will break down for the same reasons as explained in Sect. 4.2.3. When one fixes, in addition, Z_2's value to z_2, then X

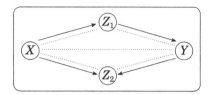

Fig. 4.5 X and Y are correlated ($X \cdots Y$) because X and Y are connected by the directed causal path $X \longrightarrow Z_1 \longrightarrow Y$. When Z_1's value is fixed, then probabilistic influence between X and Y breaks down ($INDEP_P(X, Y|Z_1)$). In that case, conditionalizing on Z_2 links X and Y up again ($DEP_P(X, Y|\{Z_1, Z_2\})$)

and Y will become linked up again (by $\{Z_1, Z_2\}$). In that case we would look at the subpopulation $D_{z_1} \cap D_{z_2}$. If z_2 is very high, for example, and the X-values are low in $D_{z_1} \cap D_{z_2}$, then these high Z_2-values z_2 can only be due to high Y-values in this subpopulation.

Before we can go on, we have to take a closer look at a special case of a causal connection by a directed or a common cause path and a common effect path of the form (d) we just discussed. So what if two variables X and Y are connected by a directed or a common cause path *and* a path of the form (d) where both paths go through one and the same variable Z? Take the causal path $X \longrightarrow Z \overset{\longleftarrow}{\longrightarrow} Y$, which consists of the two paths $X \longrightarrow Z \longrightarrow Y$ and $X \longrightarrow Z \longleftarrow Y$, as an example. In such a case X and Y would be correlated because of path $X \longrightarrow Z \longrightarrow Y$: High X-values would lead to high Z-values which would lead to high Y-values, and thus, $DEP_P(X, Y)$. But if one would fix Z's value, then this probabilistic influence would not break down as in the case of an underlying causal structure as the one depicted in Fig. 4.5. Because low X-values would indicate high Y-values in a subpopulation D_z of D with very high Z-values z, X and Y would be correlated in such a subpopulation.

Let me briefly summarize our findings so far:

- A causal path of the form (a), (b), or (c) which connects X and Y produces probabilistic dependence between X and Y. So the arrows of causal paths of the form (a), (b), or (c) behave in exactly the same way as the undirected binary causal relation we introduced in Sect. 4.2.2 to explain screening off.
- When one conditionalizes on at least one variable of every causal path of the form (a), (b), or (c) connecting X and Y and on no other variables, then X and Y become probabilistically independent.
- Common effect paths with length three ($X \longrightarrow Z \longleftarrow Y$) produce no correlation between X and Y. They do, however, also not destroy any correlation produced by a path of the form (a), (b), or (c).
- Two variables X and Y are probabilistically dependent conditional on a common effect Z lying on a path of the form $X \longrightarrow Z \longleftarrow Y$ (regardless of whether there are any paths of the form (a), (b), or (c) connecting X and Y and whether one conditionalizes on one or more variables lying on one or more of these paths).

We now know how causal paths of the form (a)–(c) as well as the special case $X \longrightarrow Z \longleftarrow Y$ of (d) probabilistically behave, and also how these different kinds of causal connections behave when they occur combined. One thing we have not investigated so far is which probabilistic dependencies and independencies more complex versions of structure (d) produce. To this end, let us take a closer look at the causal structure underlying the exemplary causal system $\langle V_e, P_e \rangle$ with $V_e = \{X, Y, Z_1, Z_2, Z_3, Z_4, Z_5, Z_6\}$ depicted in Fig. 4.6: According to our findings, subpath $Z_1 \longleftarrow Z_2 \longrightarrow Z_3$ propagates probabilistic influence from Z_1 to Z_3 and vice versa, while the two subpaths $X \longrightarrow Z_1 \longleftarrow Z_2$ and $Z_2 \longrightarrow Z_3 \longleftarrow Y$ do not so with respect to X and Z_2 as well as to Z_2 and Y, respectively. So X and Y will be probabilistically independent unconditionally ($INDEP_{P_e}(X, Y)$): In a population D, increased X-values x lead to increased Z_1-values. Since Z_1 is not a cause, but an effect of Z_2, Z_2-values will be equally distributed in subpopulation D_x of D with increased X-values as in D, and thus, no probabilistic influence of X is transported further than to Z_1. But if one would fix Z_1's value to z_1, i.e., take a look at subpopulation D_{z_1}, then low X-values would correlate with high Z_2 values in case Z_1's values are high in D_{z_1}. This would, since Z_2 is a cause of Z_3, also lead to higher Z_3-values in D_{z_1}. But, since Z_3 is not a cause of Y, Y-values will still be equally distributed among individuals u in D_{z_1} as in D. But if one would, in addition, also fix Z_3's value, i.e., look at a certain subpopulation $D_{z_1} \cap D_{z_3}$ of D, then, in case z_3 is very low, this can only be due to quite low Y-values in $D_{z_1} \cap D_{z_3}$. So X and Y should be correlated in $D_{z_1} \cap D_{z_3}$. Probabilistic dependence between X and Y should, however, break down again if we also hold the common cause Z_2's value fixed, i.e., if we take a look at subpopulation $D_{z_1} \cap D_{z_2} \cap D_{z_3}$. So a causal path between two variables X and Y of the kind (d) transports probabilistic influence from X to Y and vice versa if the values of all common effects lying on this path are fixed and all non-common effects' values (i.e., intermediate or common causes' values) are allowed to vary.

Another interesting thing is that also conditionalizing on effects of common effects may lead to linking-up. I will explain this, again, by means of our exemplary causal structure depicted in Fig. 4.6: Conditionalizing on the effect Z_6 of Z_2's and Y's common effect Z_3 will make Z_2 and Y probabilistically dependent. To see this, take a look at a subpopulation D_{z_6} of individuals with high Z_6-values z_6. Since Z_3 is the only direct cause of Z_6, these high Z_6-values can only be due to high Z_3-values. If, for example, Z_2-values are very low in D_{z_6}, then these high Z_3-values must have been produced by high Y-values, and thus, Z_2 and Y are correlated in D_{z_6}.

We can conclude that causal paths of the form (d) do not produce probabilistic dependence between two variables X and Y they connect as long as one does not conditionalize on any subset of the causal system of interest's variable set. When one conditionalizes, however, on all common effects lying on such a path or on at least one of their effects, then X and Y become probabilistically dependent. We can now try to generalize our findings and explicate our notion of causal connection required for specifying (*) above in the following way:

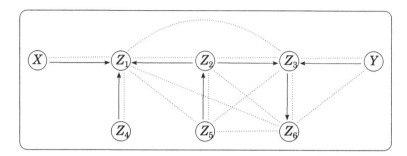

Fig. 4.6 X and Y are uncorrelated unconditionally, but become probabilistically dependent when conditionalizing on both common effects Z_1 and Z_3, i.e., on $\{Z_1, Z_3\}$. Instead of on Z_3, one can also conditionalize on the effect Z_6 of the common effect Z_3; X and Y would still be linked up by $\{Z_1, Z_6\}$. Probabilistic influence between X and Y, however, breaks down when also conditionalizing on the common cause Z_2, i.e., on $\{Z_1, Z_2, Z_3\}$ or $\{Z_1, Z_2, Z_6\}$. Neither conditionalizing on the cause Z_4 of the common effect Z_1, nor conditionalizing on the cause Z_5 of the common cause Z_2 breaks down the probabilistic dependence between X and Y established via conditionalizing on $\{Z_1, Z_3\}$ or $\{Z_1, Z_6\}$. Dotted lines ('$\cdots$') stand for unconditional correlations and arrows ('$\longrightarrow$') for direct causal relations between two variables

Definition 4.4 (causal connection) X and Y (with $X, Y \in V$) are causally connected by $M \subseteq V\backslash\{X, Y\}$ in a causal structure $\langle V, E \rangle$ if and only if X and Y are connected via a causal path π in $\langle V, E \rangle$ such that all colliders (or effects of colliders) lying on this path π are in M and no intermediate or common cause lying on this path π is in M.

The above explicated notion of causal connection is essentially identical to the causal reading of d-connection in the d-connection condition (Condition 3.2).

Let us now implement this notion of causal connection into (*). When doing so, we arrive at the following (still preliminary) axiom:

Preliminary Axiom 4.4 *If $\langle V_e, P_e \rangle$ is an empirical system, then there is a causal structure $\langle V_e, E \rangle$ such that for all $X, Y \in V_e$ and $M \subseteq V_e\backslash\{X, Y\}$: $DEP_{P_e}(X, Y | M)$ if and only if X and Y are causally connected (in the sense of Definition 4.4) given M in $\langle V_e, E \rangle$.*

This (preliminary) axiom provides an explanation of the probabilistic dependencies and independencies implied by an empirical system's associated probability distribution by referring to this empirical system's underlying causal structure (which produces these dependencies and independencies). It is, however, still deficient. To see this, let us take the barometer scenario discussed in Sect. 4.2.2 as an exemplary case again. The barometer scenario is an empirical system $\langle V_e, P_e \rangle$ with $V_e = \{B, P, S\}$, where B stands for barometer reading, P for atmospheric pressure, and S for the rising of a storm. Let us take a closer look at another empirical system now: the empirical system $\langle V_e', P_e' \rangle$ with $V_e' = \{B, S\}$ and $P_e' = P_e \uparrow V_e'$, which is a subsystem of the barometer scenario. We will find that B and S are correlated in this empirical system's probability distribution P_e'. According to Preliminary Axiom 4.4, this correlation can only be due to a causal connection, and thus, B must be a direct cause of S or S must be a direct cause of B. But neither the one nor the other is true: B and S are not correlated because one causes the other, but because they have a common cause, *viz.* the atmospheric pressure P. The problem is that this common cause P is not captured by the empirical system's variable set V'_e. This problem can be solved when the causal structures Preliminary Axiom 4.4 postulates to explain probabilistic dependencies and independencies implied by an empirical system's probability distribution are allowed to contain more variables than this empirical system does.

To modify Preliminary Axiom 4.4 in the proposed way the notion of a causal model (see Sect. 3.2) is required. Note that not every causal system $\langle V, E, P \rangle$ is capable of producing the probabilistic dependencies and independencies implied by an empirical system $\langle V_e, P_e \rangle$'s probability distribution P_e. First of all, V_e must be a subset of V. Secondly, P_e should be the restriction of P to V_e ($P \uparrow V_e = P_e$). And last but not least, the causal system's causal structure $\langle V, E \rangle$ has to imply the correct probabilistic dependencies and independencies. This means, as we have seen so far, that for every correlation $DEP_P(X, Y|M)$ there must be a causal connection between X and Y given M in $\langle V, E \rangle$, while this causal structure features no causal connection between X and Y given M for every $INDEP_P(X, Y|M)$. If a causal structure satisfies all three conditions with respect to a given empirical system, then we say that this causal system *robustly produces* or *robustly underlies* the corresponding empirical system[4]:

Definition 4.5 (robustly underlying causal system) If $\langle V_e, P_e \rangle$ is an empirical system and $\langle V, E, P \rangle$ is a causal system, then $\langle V, E, P \rangle$ robustly produces (or robustly underlies) $\langle V_e, P_e \rangle$ if and only if

(a) $V_e \subseteq V$, and
(b) $P \uparrow V_e = P_e$, and
(c) for all $X, Y \in V$ and $M \subseteq V \backslash \{X, Y\}$: $DEP(X, Y|M)$ if and only if X and Y are causally connected (in the sense of Definition 4.4) given M in $\langle V, E \rangle$.

For every empirical system there may be more than one robustly underlying causal system, and, vice versa, every causal system typically robustly produces more than just one empirical system. What Definition 4.5 ultimately does is to distinguish between causal structures that are possible candidates to explain the probabilistic dependence and independence relations implied by an empirical system's probability distribution and causal structures that are not. Figures 4.7 and 4.8 show some examples of empirical systems and some causal systems which would robustly produce these empirical systems.

---

[4]I call causal systems satisfying the three mentioned conditions "robustly" producing (or underlying) the corresponding empirical systems since these causal systems presuppose robust (or faithful) probability distributions. In the next subsection, I will also introduce underlying systems for explaining (in)dependencies in empirical systems without robust probability distributions.

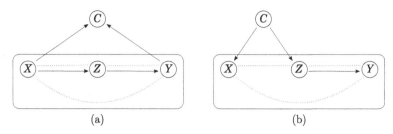

Fig. 4.7 The boxes in (**a**) and (**b**) stand for an empirical system $\langle V_e, P_e \rangle$'s variable set V_e. The dotted lines ('$\cdots$') stand for unconditional probabilistic dependencies in P_e. X and Y are screened off by Z in P_e ($INDEP_{P_e}(X, Y|Z)$). (**a**) A causal system with causal structure $X \longrightarrow Z \longrightarrow Y$ as well as one with causal structure $C \longleftarrow X \longrightarrow Z \longrightarrow Y \longrightarrow C$ could explain $\langle V_e, P_e \rangle$. In the first case, this causal system's variable set would be V_e and its probability distribution P_e. (**b**) A causal system with causal structure $X \longleftarrow C \longrightarrow Z \longrightarrow Y$ could also explain $\langle V_e, P_e \rangle$. Both causal systems would produce $DEP_{P_e}(X, Y)$ and $INDEP_{P_e}(X, Y|Z)$ in P_e

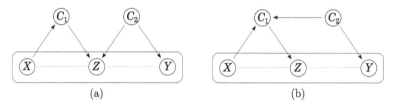

Fig. 4.8 The boxes in (**a**) and (**b**) stand for an empirical system $\langle V_e, P_e \rangle$'s variable set V_e. The dotted lines ('$\cdots$') stand for unconditional probabilistic dependencies in P_e. X and Y are linked up by Z in P_e ($DEP_{P_e}(X, Y|Z)$). In (**a**), X and Y become probabilistically dependent when conditionalizing on Z because Z is the only collider on a collider path connecting X and Y. In (**b**), X and Y are linked up by Z because Z is an effect of a collider lying on a collider path connecting X and Y

With help of the notion of a robustly underlying causal system, Preliminary Axiom 4.4 can now be generalized to the following (preliminary) axiom:

Preliminary Axiom 4.5 *If $\langle V_e, P_e \rangle$ is an empirical system, then there is a causal system $\langle V, E, P \rangle$ robustly producing $\langle V_e, P_e \rangle$ (in the sense of Definition 4.5).*

Preliminary Axiom 4.5 states that for every empirical system there is a causal system which robustly produces this empirical system. In other words, the (preliminary) axiom says that the probabilistic dependencies and independencies implied by an empirical system's associated probability distribution are explainable by some causal system satisfying conditions (a)–(c) of Definition 4.5.

Note that Preliminary Axiom 4.5 basically asserts that for every empirical system there is a (maybe larger) causal system such that for every (conditional) dependence there is a corresponding causal connection in this system, and, vice versa, that features a (conditional) dependence for every causal connection. The former amounts to assuming CMC (or the *d*-connection condition) for this underlying causal system, the latter to assuming the full faithfulness condition CFC. That also

CFC is presupposed is mainly due to the argumentation throughout Sect. 4.2 so far, in which I assumed that our empirical systems' probability distributions are robust. In the next subsection I will present a weaker version of Preliminary Axiom 4.5 which also allows for explaining screening off and linking-up in empirical systems with non-robust probability distributions. This weaker version will only presuppose CMC to hold.

4.2.5 Explaining Screening Off and Linking-Up in Systems Without Robust Probability Distributions

In this subsection I show how the explanation of screening off and linking-up developed in Sect. 4.2.4 can be generalized to empirical systems with non-robust probability distributions (see also Schurz and Gebharter 2016, sec. 2.2). If a probability distribution P_e of an empirical system $\langle V_e, P_e \rangle$ is non-robust, i.e., does feature non-faithful independencies which would be destroyed by arbitrary small parameter changes, then also a system $\langle V, E, P \rangle$ with $V \supseteq V_e$ and $P \uparrow V_e = P_e$ will be non-faithful. So if we have to deal with an empirical system $\langle V_e, P_e \rangle$ featuring non-faithful independencies, we will not find a causal system robustly producing P_e in the sense of Definition 4.5. Put differently: Empirical systems with non-robust probability distributions can only be explained by (maybe larger) non-faithful causal systems. The first thing to do to account for this fact is to modify the notion of a (robustly) underlying causal system Definition 4.5:

Definition 4.6 (underlying causal system) If $\langle V_e, P_e \rangle$ is an empirical system and $\langle V, E, P \rangle$ is a causal system, then $\langle V, E, P \rangle$ produces (or underlies) $\langle V_e, P_e \rangle$ if and only if

(a) $V_e \subseteq V$, and
(b) $P \uparrow V_e = P_e$, and
(c) for all $X, Y \in V$ and $M \subseteq V \backslash \{X, Y\}$: If $DEP(X, Y|M)$, then X and Y are causally connected (in the sense of Definition 4.4) given M in $\langle V, E \rangle$.

The only difference between Definitions 4.6 and 4.5 is condition (c): While the latter requires the underlying system to be Markov and faithful, the former only requires it to be Markov.

With this slight modification at hand, we can now formulate the following *causal Markov axiom* (CMA):

Axiom 4.1 (causal Markov axiom) *If $\langle V_e, P_e \rangle$ is an empirical system, then there is a causal system $\langle V, E, P \rangle$ producing $\langle V_e, P_e \rangle$ (in the sense of Definition 4.6).*

Axiom 4.1 differs from the old Preliminary Axiom 4.5 only in so far as it uses a weaker notion of production. Axiom 4.1 claims that every (conditional or unconditional) dependence in an empirical system is produced by some (maybe larger) causal structure. Axiom 4.1 is intended to hold for empirical systems with as

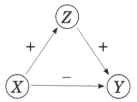

Fig. 4.9 The system's parameters are assumed to be fine-tuned in such a way that paths $X \longrightarrow Y$ and $X \longrightarrow Z \longrightarrow Y$ cancel each other exactly. In that case, the structure depicted above can explain why X and Y are linked up by Z

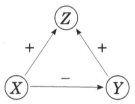

Fig. 4.10 The system's parameters are assumed to be fine-tuned in such a way that paths $X \longrightarrow Y$ and $X \longrightarrow Z \longleftarrow Y$ cancel each other exactly when $X \longrightarrow Z \longleftarrow Y$ is activated. In that case, the structure depicted above can explain why X and Y are screened off by Z

well as without robust probability distributions. In case of empirical systems with robust distributions, screening off and linking-up are explained exactly as illustrated in Sect. 4.2.4. One gets the implication from causal connection (or d-connection) to dependence for the variables of the empirical system missing in Axiom 4.1 for free due to the robustness of the system's probability distribution P_e.

If our system $\langle V_e, P_e \rangle$'s probability distribution is not robust, then Axiom 4.1 allows for new combinations of screening off and linking-up phenomena and causal structures explaining them. Assume, for example, we are interested in an empirical system $\langle V_e, P_e \rangle$ with $V_e = \{X, Y, Z\}$ in which Z links up X and Y. If P_e is not robust, then linking-up in $\langle V_e, P_e \rangle$ can, for instance, be explained by a non-faithful causal system $\langle V, E, P \rangle$ with the causal structure depicted in Fig. 4.9. Here paths $X \longrightarrow Y$ and $X \longrightarrow Z \longrightarrow Y$ are assumed to cancel each other exactly. But X's causal influence on Y over path $X \longrightarrow Y$ can be isolated by conditionalizing on Z, which blocks path $X \longrightarrow Z \longrightarrow Y$ and reveals Y's dependence on X. Thus, we will find that $INDEP_P(X, Y)$ and $DEP_P(X, Y|Z)$.

For another example of a new possibility to explain screening off in empirical systems without robust probability distributions, assume our system is $\langle V_e, P_e \rangle$ with $V_e = \{X, Y, Z\}$ in which Z screens X off from Y. This can be explained, for example, by a causal system $\langle V, E, P \rangle$ with the causal structure depicted in Fig. 4.10, where path $X \longrightarrow Z \longleftarrow Y$ is assumed to exactly cancel $X \longrightarrow Y$ when activated. Thus, we will find $DEP_P(X, Y)$ and $INDEP_P(X, Y|Z)$.

Note that for explaining non-robust screening off and linking-up one requires, strictly speaking, not only that the underlying system satisfies CMC, but also that all causal paths involved in bringing about the screening off or linking-up phenomenon are productive. For details, see Schurz and Gebharter (2016, sec. 2.2).

Summarizing, driven by an inference to the best explanation for screening off and linking-up phenomena, we were able to develop CMA on purely empirical grounds. We only put into CMA what we had to in order to account for these empirical phenomena. So we finally arrive at an axiomatization of the causal Bayes net framework consisting in CMA (Axiom 4.1), CPA (Axiom 3.1), and RCFA (Axiom 3.2). My claim is that the resulting theory of causation can be seen as a real alternative to more classical theories of causation. In addition, its core axiom CMA can be justified by an inference to the best explanation of statistical phenomena. In the next section I will show that certain versions of the theory also provably lead to empirical consequences. The theory behaves, thus, exactly like a (good) empirical theory of the sciences, which seems to be something novel and desirable for a philosophical theory.

4.3 Empirical Content of the Theory of Causal Nets

4.3.1 Empirical Content of the Causal Markov Axiom

The typical objections to characterizing causation by means of the causal Markov condition (CMC)—which is exactly what CMA does—consist in constructing causal models that violate CMC. The most prominent of these objections is maybe Cartwright's (1999a; 2007, p. 107) chemical factory (recall Sect. 3.3). Proponents of assuming CMC typically counter such arguments by claiming that CMC's opponents cannot guarantee that their counterexamples do not misrepresent causal processes or omit causal arrows or common causes. They point the opponents of CMC to the alleged fact that almost all known empirical systems in our world satisfy CMC (cf. Hitchcock 2010, sec. 3.3; Pearl 2000, pp. 62f; Spirtes et al. 2000, p. 29). In this subsection I want to present a result that is technically well-known, but whose philosophical consequences seem to have been left unconsidered in the discussion:

Theorem 4.3 (Schurz and Gebharter 2016, theorem 3) *Every analytically possible empirical model $\langle V_e, P_e \rangle$ can be expanded to an acyclic causal model $\langle V, E, P \rangle$ satisfying CMC and Prod.*

This theorem basically shows that even a stronger version of CMA is empirically empty: One gets no empirical content, even if one assumes that the causal system underlying the empirical system of interest is acyclic and satisfies Prod in addition to CMC. So, strictly speaking, there cannot exist counterexamples to CMA. The result's philosophically interesting implication is that CMA can neither be falsified by providing counterexamples, nor confirmed by the fact that CMC is satisfied by almost all known empirical systems. What can be shown by counterexamples is only that the conjunction of CMA and additional causal assumptions is violated.

But this does not have to worry proponents of CMA too much. As shown in Sect. 4.2, there are strong reasons for CMA that do not require that CMA can be empirically confirmed: CMC-satisfying causal models provide the best explanation for screening off and linking-up phenomena in empirical systems.

4.3.2 Empirical Content of Assuming Faithfulness

Recall from Sect. 3.4.2 that the full causal faithfulness condition (CFC) may be violated. However, even assuming the full faithfulness condition (alone) for all kinds of systems would not imply any empirical consequences. This is our next theorem:

Theorem 4.4 *Every analytically possible empirical model $\langle V_e, P_e \rangle$ can be expanded to an acyclic causal model $\langle V, E, P \rangle$ satisfying CFC.*

Proof We show that for every empirical model $\langle V_e, P_e \rangle$ there is an acyclic expansion $\langle V_e, E, P_e \rangle$ that satisfies CFC. The faithfulness condition is satisfied by a causal model $\langle V_e, E, P_e \rangle$ if and only if every d-connection between two variables $X, Y \in V_e$ given a subset $M \in V_e \backslash \{X, Y\}$ produces a probabilistic dependence $DEP_{P_e}(X, Y|M)$. So the faithfulness condition is trivially satisfied if $E = \emptyset$. Moreover, $\langle V_e, E, P_e \rangle$ is an acyclic causal model.   □

This result implies that also postulating an axiom assuming any weaker version of the faithfulness condition (such as in RCMA, Zhang and Spirtes' 2008, p. 253 adjacency- and orientation-faithfulness conditions, or the productivity condition) could not be empirically tested without further assumptions. For an overview of weaker faithfulness conditions, see Zhang and Spirtes (2016).

4.3.3 Empirical Content of the Causal Markov Axiom and Assuming Faithfulness

If one would assume both CMA and CFC, the resulting version of the theory of causal nets would, for the first time, lead to empirically testable consequences. CMA together with assuming faithfulness excludes analytically possible empirical models of the following kind:

Theorem 4.5 (Schurz and Gebharter 2016, theorem 4) *No empirical model $\langle V_e, P_e \rangle$ with $X, Y, Z \in V_e$ verifying the logically possible (in)dependence relations in (a) or (b) can be expanded to a causal model $\langle V, E, P \rangle$ satisfying CMC and CFC:*

(a) *(i) $DEP_{P_e}(X, Z|M) \wedge DEP_{P_e}(Z, Y|M)$ holds for all $M \subseteq V_e \backslash \{X, Y\}$, and (ii) there exist two distinct sets $N, N' \subseteq V_e \backslash \{X, Y\}$ with $Z \in N$ but $Z \notin N'$ which screen off X fom Y.*

(b) $INDEP_{P_e}(X, Y)$, $INDEP_{P_e}(Y, Z)$, $INDEP_{P_e}(X, Z)$, $DEP_{P_e}(X, Y|Z)$, DEP_{P_e} $(Y, Z|X)$, $DEP_{P_e}(X, Z|Y)$.

Recall from Sect. 3.4.2 that assuming that all causal systems satisfy the full faithfulness condition is a much too strong claim that cannot be justified by assuming that most causal systems' parameters are modulated by external noise. However, the external noise assumption can be used to justify the restricted causal faithfulness axiom (Axiom 3.2). Hence, RCFA together with CMA generates probabilistic empirical content for the resulting version of the theory of causal nets:

Corollary 4.1 (Schurz and Gebharter 2016, Corollary 1) *CMA and RCFA imply that analytically possible empirical models* $\langle V_e, P_e \rangle$ *which satisfy conditions (a) and (b) of Axiom 3.2 and conditions (a) or (b) of Theorem 4.5 are highly improbable.*

4.3.4 Empirical Content of Adding Time Directedness

In this subsection I present results about how the theory of causal nets' empirical content increases when assuming that causes always precede their effects in time. To make this assumption explicit, we have to modify the notion of a causal model. In the whole subsection 'causal model' will stand short for 'causal event model'. A *causal event model* is a quadruple $\langle V, E, P, t \rangle$. V's elements are event variables and $t : V \to \mathbb{R}$ is a time function. $t(X)$ stands for the time point at which the events described by X, i.e., X's instantiations to its possible values, occur.

With help of this notion of an event model, we can now formulate the following *axiom of time-directedness* (Temp) (cf. Schurz and Gebharter 2016, p. 1096):

Axiom 4.2 (time-directedness) *For every causal event model* $\langle V, E, P, t \rangle$ *the following holds: If* $X \longrightarrow Y$ *in* $\langle V, E \rangle$, *then* $t(X) < t(Y)$.

In the light of CMA and the full faithfulness condition CFC, assuming Temp excludes screening off of two variables X and Y by a future cause Z as well as linking-up of two variables X and Y by a common cause Z that lies in the past of at least one of these two events X or Y. The causal screening off and linking-up scenarios rendered possible and impossible by CMA, CFC, and Temp are depicted in Fig. 4.11.

These findings can be generalized: While CMA together with assuming CFC and Temp excludes screening off by future events in general (see Theorem 4.6(a) below), they exclude linking-up by events lying in the past of at least one of the linked up events only under the proviso that the event Z linking up X and Y is not screened off from X and Y by past events (see Theorem 4.6(b) below). For an example of a causal scenario in which this latter requirement is not met, see Fig. 4.12.

Theorem 4.6 (Schurz and Gebharter 2016, theorem 5) *No empirical event model* $\langle V_e, P_e, t \rangle$ *featuring the following combinations of probabilistic dependencies and independencies can be expanded to a causal model* $\langle V, E, P, t' \rangle$ *satisfying CMC, CFC, and Temp:*

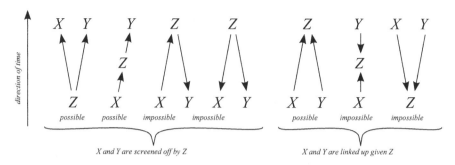

X and Y are screened off by Z   X and Y are linked up given Z

Fig. 4.11 Causal screening off and linking-up scenarios rendered possible/impossible by CMA together with assuming CMC and Temp

Fig. 4.12 In case a variable Z lying in the past of X or Y is screened off from X and Y by past common causes C_i, Z can link up X and Y

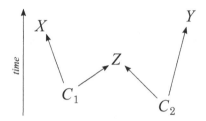

(a) $DEP_{P_e}(X, Y)$ and $INDEP_{P_e}(X, Y|Z)$ such that $t(Z) > t(X)$ and $t(Z) > t(Y)$ hold.

(b) $INDEP_{P_e}(X, Y)$ and $DEP_{P_e}(X, Y|Z)$, where the following holds for ψ (where $\psi = X$ or $\psi = Y$): $t(Z) < t(\psi)$ and there exists no C with $t(C) < t(Z)$ and $INDEP_{P_e}(\psi, Z|C)$.

Theorem 4.6 above makes use of the full faithfulness condition CFC, which may be considered as problematic, since CFC cannot be justified by the external noise assumption. But recall that the restricted causal faithfulness axiom RCFA is fully justified by assuming that (almost all) causal systems in our world are influenced by external noise. Thus, we can also derive a theorem that says that causal models which satisfy one of the two conditions (a) or (b) in Theorem 4.6 are highly improbable in the light of CMA, Temp, and RCFA.

If we compare our results so far with Reichenbach's (1956/1991, p. 162) analysis of the direction of time on the basis of causal relations, we find that though Reichenbach made heavy use of screening off relationships to account for causal connections, he totally overlooked the connection between causal structures and linking-up properties. Accordingly, Reichenbach did not assume any kind of faithfulness. What Reichenbach showed (without requiring any kind of faithfulness) is that two simultaneous event variables X and Y cannot be screened off by a future event variable Z without there being a common cause Z' lying in the past of X and Y. The following theorem shows that Reichenbach's result can be reproduced within the theory of causal nets:

Theorem 4.7 (Schurz and Gebharter 2016, theorem 6.1) *No empirical event model $\langle V_e, P_e, t \rangle$ with two simultaneous event variables $X, Y \in V$ such that $DEP_{P_e}(X, Y)$ and $INDEP_{P_e}(X, Y|Z)$ hold for some Z with $t(Z) > t(X) = t(Y)$ can be expanded to a causal event model $\langle V, E, P, t' \rangle$ satisfying CMC and Temp without a set M of event variables lying in the past of X and Y such that $INDEP_P(X, Y|M)$ holds.*

The next theorem shows that it is highly improbable that there exists a future event variable Z screening off two event variables X and Y without a deterministic dependence of X, of Y, or of some set M of event variables lying in the past of X and Y that screens off X from Y on Z. This result requires RCFA in addition to CMA and Temp:

Theorem 4.8 (Schurz and Gebharter 2016, theorem 6.2) *In the light of Temp and RCFA it is highly improbable that an empirical event model $\langle V_e, P_e, t \rangle$ with two simultaneous event variables $X, Y \in V$ such that $DEP_{P_e}(X, Y)$ can be expanded to a causal event model $\langle V, E, P, t' \rangle$ satisfying CMC that features a variable Z with $t(Z) > t(X) = t(Y)$ screening off X and Y such that neither X, nor Y or some set M of variables lying in the past of X and Y that screens off X from Y depends deterministically on some Z-value z.*

One can also add an axiom of locality. The simplest version of such an axiom could state that a direct causal connection $X \longrightarrow Y$ implies that X-events lie within the backward light cones of Y-events. For the consequences of adding such an axiom, see Schurz (2015, p. 91).

4.3.5 Empirical Content of Adding Intervention Assumptions

In this subsection I show that adding an axiom stating the independence of (almost all) human actions from certain variables of the causal systems we are manipulating to CMA leads to empirical content, even in the absence of any kind of faithfulness or minimality assumption. We formulate the following *axiom of the independence of human interventions* (IHI) (cf. Schurz and Gebharter 2016, p. 1098):

Axiom 4.3 (independence of human interventions) *Most of a person's actions $I = i$ manipulating a variable of a person-external causal system $\langle V, E, P \rangle$ that are experienced as "free" are probabilistically independent of those variables in V that are non-effects of I.*

Note that the notion of "freedom" used in Axiom 4.3 is extremely weak. According to this notion, a human action $I = i$ is, for example, free w.r.t. a person-external causal system $\langle V, E, P \rangle$ if there are no common causes Z of I and non-effects $X \in V$ of I. But even if there are such common causes Z of I and non-effects $X \in V$, $I = i$ can still be a free action w.r.t. $\langle V, E, P \rangle$. It suffices that all common causes Z are fixed to a certain value z in the background context $C = c$. This will screen I off from X.

In the literature, interventions are typically modeled by so-called intervention variables I. For I to be an intervention variable for another variable X one typically assumes that I is an exogenous variable that is a direct cause of X and only of X, that is independent of all of its non-effects in the system of interest, and that has at least some probabilistic influence on X (cf. Spirtes et al. 2000, sec. 3.7.2; Eberhardt and Scheines 2007, pp. 984–987). An intervention on X is then defined as an intervention variable I for X taking some value i. Alternatively, one can define a three-place notion of an intervention variable such as Woodward (2003, p. 98) prefers. The idea is basically the following: We define an intervention variable I for X w.r.t. a test variable Y as a variable I that causes X, that is independent of all non-effects of I in the system of interest, and that causes Y (if it causes Y at all) only over causal paths going through X. Also here an intervention (on X w.r.t. Y) is then defined as a value realization of an intervention variable. In this subsection, we will use the two-place notion of an intervention variable and not some kind of Woodwardian three-place notion. (For details on Woodward's notion of intervention, see Chap. 5.) I use an upper-case 'I' for an intervention variable and write the intervention variable's target variable as an index. Thus, I_X is an intervention variable for X, I_Y is an intervention variable for Y, etc.

Interventions are typically not required to be human actions; also natural phenomena that satisfy the conditions mentioned above can be regarded as interventions. However, Axiom 4.3 nicely connects human actions to interventions. It states that (almost) all human actions which are experienced as "free" and whose targets are person-external causal systems qualify as interventions in the sense explained above. Thus, adding IHI to CMA and assuming acyclicity leads to empirical content, as the following theorem shows:

Theorem 4.9 *No empirical model* $\langle V_e, P_e \rangle$ *with* $X, Y \in V_e$ *such that* X *and* Y *can be independently manipulated by free human actions* $I_X = i_X$ *and* $I_Y = i_Y$ *that features dependencies* $DEP_{P_e}(I_X, Y)$ *and* $DEP_{P_e}(I_Y, X)$ *can be expanded to an acyclic causal model* $\langle V, E, P \rangle$ *satisfying CMC.*

Proof Assume $\langle V_e, P_e \rangle$ is an empirical model with $X, Y \in V_e$ and intervention variables I_X and I_Y for X and Y, respectively, such that $DEP_{P_e}(I_X, Y)$ and $DEP_{P_e}(I_Y, X)$ hold. Now assume that $\langle V, E, P \rangle$ expands $\langle V_e, P_e \rangle$ and that $\langle V, E, P \rangle$ is acyclic and satisfies CMC. We show that $DEP_P(I_X, Y)$ and $DEP_P(I_Y, X)$ together lead to a causal cycle $X \xrightarrow{\leftarrow} \xrightarrow{\leftarrow} Y$ in $\langle V, E \rangle$, what contradicts the assumption of acyclicity.

From $DEP_P(I_X, Y)$ and the d-connection condition it follows that I_X and Y are d-connected (given the empty set) in $\langle V, E \rangle$. Thus, there is a path $\pi : I_X - \ldots - Y$ not featuring a collider. Since I_X is an intervention variable for X, I_X is exogenous and a direct cause of X and only of X. Hence, π must have the form $\pi : I_X \longrightarrow X - \ldots - Y$. Since π does not feature colliders, it must have the form $\pi : I_X \longrightarrow X \longrightarrow \longrightarrow Y$.

In exactly the same way as it follows from $DEP_P(I_X, Y)$ that there has to be a path $\pi : I_X \longrightarrow X \longrightarrow\longrightarrow Y$, it follows from $DEP_P(I_Y, X)$ that there must be a path $\pi^* : I_Y \longrightarrow Y \longrightarrow\longrightarrow X$. Thus, $\pi^{**} : I_X \longrightarrow X \underset{\longrightarrow}{\overset{\longleftarrow}{}} Y \longleftarrow I_Y$, which is the concatenation of π and π^*, will be part of $\langle V, E, P \rangle$'s causal structure. But this contradicts the initial assumption of acyclicity. $\square$

4.4 Conclusion

This chapter started with Hume's skeptical challenge (see Sect. 4.1): Are there good reasons to belief in causation as something ontologically real out there in the world, or is causality merely a subjective feature of our minds, while only regularities (i.e., correlations) are real? The basic idea underlying this chapter is that causation behaves exactly like a theoretical concept. This means that causation cannot be explicitly defined, but only implicitly characterized by axioms which connect causal structures to observable phenomena. The only difference between causation and theoretical concepts, such as force, is that the theory of causation does not belong to a specific discipline (such as force, which belongs to physics), but to an interdisciplinary theory. Causation, as characterized within the theory of causal nets, satisfies two important standards for theoretical concepts: They are introduced and justified by an inference to the best explanation, and the theory containing these concepts is empirically testable.

In Sect. 4.2.1 we found that the observation of mere correlations is insufficient to justify the assumption of causal relations. Postulating a causal relation for every correlation is a mere duplication of entities. Thus, this justification attempt falls victim to Occam's razor. However, there are more complex statistical phenomena whose explanation requires the assumption of more complex causal structures: screening off and linking-up. These phenomena can best be explained by structures consisting of directed binary causal relations which transport probabilistic dependencies. If the probability distributions of the systems we are interested in are robust, the observation of screening off and linking-up phenomena forces us to assume CMC and CFC. In case the distributions of the systems of interest are not robust or we do not know whether they are robust, we have only to assume CMC (plus productive causal paths). This can be seen as strong support for causation (as characterized within the theory of causal nets) as something ontologically real.

In Sect. 4.3 I presented results obtained in Schurz and Gebharter (2016) and also new results concerning the theory of causal nets' empirical content. We found in Sects. 4.3.1 and 4.3.2 that neither CMA nor assuming the full causal faithfulness condition alone excludes any logically possible probability distribution. We also found that CMA together with CPA and the assumption of acyclicity is empirically empty. Though the technical result that for every empirical model $\langle V_e, P_e \rangle$ there could be constructed a causal model $\langle V_e, E, P_e \rangle$ satisfying CMC is not new, its philosophical implications seem to have been ignored in the literature. Constructing causal models that do not satisfy CMC can, strictly speaking, only show that certain

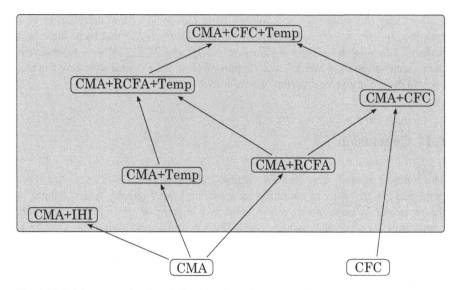

Fig. 4.13 DAG representing the relationship of certain versions of the theory of causal nets w.r.t. their empirical content. Arrows '$X \longrightarrow Y$' indicate that X's empirical consequences are a proper subset of Y's empirical consequences such that there is no Z whose empirical consequences are a proper subset of Y and a proper superset of X. Theory versions within the grey box do have empirical consequences; versions not in the box are empirically empty

causal models cannot correctly represent the systems' underlying causal structure. To falsify the theory as a whole (or, more accurately, a version of the theory), one has to show that there are certain empirical systems that cannot be expanded to a causal model satisfying the conjunction of the theory's axioms.

CMA together with assuming the full faithfulness condition CFC actually excludes certain logically possible probability distributions (Sect. 4.3.3). Also a theory version with CMA and the weaker axiom of restricted causal faithfulness (RCFA) as axioms leads to (probabilistic) empirical content. In Sect. 4.3.4 we saw that adding the axiom of time-directedness (Temp) to different versions of the theory of causal nets also excludes certain empirical models. CMA+Temp and CMA+RCFA+Temp as well as CMA+CFC+Temp do have increasingly empirical consequences. In Sect. 4.3.5 we finally investigated the consequences of adding the axiom of the independence of human interventions (IHI) to CMA. It could be shown that CMA together with intervention assumptions also leads to empirical content. The findings described above are summarized and graphically illustrated by the DAG in Fig. 4.13.

My hope is that the axiomatization of the causal Bayes net framework developed in Chaps. 3 and 4 and the results obtained in Chap. 4 show that the theory of causal nets that results from this axiomatization is a real competitor to more classical theories of causation that comes with many advantages. It is a metaphysical lightweight and causation, as characterized within the theory, satisfies modern standards for theoretical concepts of successful empirical theories of the sciences: The theory's core axiom, CMA, can be justified by an inference to the best explanation of empirical phenomena and certain versions of the theory are empirically testable. The next two big chapters of this book are more application oriented. They use the theory of causal Bayes nets to highlight several problems with Woodward's (2003) interventionist theory of causation and shed some light on the recent debate about mechanisms within the philosophy of science. This should provide some evidence that the theory can be fruitfully used to investigate more philosophical problems in an empirically informed way.

Chapter 5
Causal Nets and Woodwardian Interventionism

Abstract In this chapter I develop a novel reconstruction of Woodward's interventionist theory of causation within the theory of causal nets. This endeavor allows one to see in which respects the two theories agree and in which respects they diverge from each other. It also allows for uncovering several weak points of Woodward's theory which may have been overlooked otherwise. I highlight some of these weak points of Woodward's interventionist theory of causation and suggest several modifications of the theory to avoid them. This results in two alternative versions of an interventionist theory. The basic causal notions of these alternative versions turn out to coincide with the corresponding causal notions within the theory of causal nets under suitable conditions.

5.1 Introduction

Woodward's (2003) interventionist theory of causation is one of the most prominent theories of causation at the moment. It has a huge amount of supporters and is even applied to problems outside research fields that are mainly focusing on causation.[1] One of the reasons for the theory's popularity may be that it is so close to very familiar intuitions about the relationship of causation and manipulation. One often can control the effect by manipulating the cause. However, the theory is, again and again, applied without giving much care to its definitorial details, what regularly leads to false consequences and/or confused discussions (cf. Baumgartner 2010 or Baumgartner and Gebharter 2016). The reason for this may, again, be the theory's closeness to intuitions: People may have the feeling that the formal details of the theory are well understood, even without taking a closer look at them. Another reason may be that Woodward himself is sometimes not very clear in spelling out these details.

---

[1]See, for instance, the reductionism debate (e.g., Eronen 2011, 2012; Fazekas and Kertesz 2011; Glennan 1996; Kistler 2009; Soom 2011) or the discussion about constitutive relevance relations in mechanistic philosophy of science (e.g., Craver 2007a,b; Craver and Bechtel 2007; Glauer 2012; Kaplan 2012).

© Springer International Publishing AG 2017
A. Gebharter, *Causal Nets, Interventionism, and Mechanisms,*
Synthese Library 381, DOI 10.1007/978-3-319-49908-6_5

One of the main goals of this chapter is to give cause to distrust in closeness to intuitions as a criterion for a good theory. Sometimes a formally more complex theory more distant from intuitions may be better than competing theories nicely matching these intuitions. In this chapter I demonstrate that Woodward's (2003) theory of causation, though it is so close to intuitions, has to face some serious problems, and that, if one wants to account for these problems, i.e., if one wants to modify Woodward's theory in such a way that it can deal with the problematic cases, it is closely related to the causal nets approach. It will turn out that unproblematic versions of Woodward's theory of causation can be shown to be consequences of assuming that certain combinations of the causal nets approach's core conditions hold.

The chapter is structured as follows: I start by giving a presentation of the core definitions of Woodward's (2003) interventionist theory of causation in Sect. 5.2.1. In particular, these are the definitions of total causation, direct causation, contributing causation, and intervention variable. I will also briefly discuss Woodward's notion of a possible intervention. In particular, I will clarify the question in which sense interventions can and should be possible. I will argue that possible interventions should not be understood as logically possible, as Woodward suggests, but as physically possible. I then provide a precise reconstruction of Woodward's core definitions in terms of the causal nets approach in Sect. 5.2.2. Everything I will do in later sections hinges on this particular reconstruction. This reconstruction is based on a reconstruction of Woodward's notion of direct causation developed in Gebharter and Schurz (2014). It differs from the latter insofar as it is based on the notion of a dependence model, i.e., an abstraction from concrete numerical details of a certain probability distribution that only represents dependencies and independencies between the variables of interest.

Next (Sect. 5.3), I will present three kinds of hypothetically possible causal scenarios. These scenarios are such that the application of Woodward's (2003) theory to them generates false results. I will highlight the conditions within Woodward's core definitions that give rise to these problems and will suggest certain modifications one could make to the theory to avoid the problems. In particular, I will discuss systems one cannot intervene on in the sense of Woodard in Sect. 5.3.1, causal chains in which intermediate causes turn out to be intervention variables in Sect. 5.3.2, and deterministic causal chains as well as deterministic common cause structures in Sect. 5.3.3. I will discuss several ways to solve these problems.

In Sect. 5.4 I will finally suggest alternative versions of Woodward's original interventionist theory. In particular, I will present one alternative version using only strong deterministic interventions and another one that uses weak stochastic interventions. While the first one is closer to Woodward's theory, the latter is also applicable to systems including variables that cannot be controlled by deterministic interventions. Both alternative theories will be the result of accepting the most reasonable modifications for deterministic and stochastic interventions, respectively, discussed in Sect. 5.3. As a last step, I will show that both unproblematic versions of Woodward's theory do not have to be seen as independent (or "stand alone") theories of causation, but turn out to be consequences of certain combinations of the causal Markov condition (CMC), the causal minimality condition (Min), and the causal faithfulness condition (CFC).

5.2   Woodward's Interventionist Theory of Causation

5.2.1   Causation and Possible Interventions

In this section I present Woodward's (2003) interventionist theory of causation, which is an advancement of agency theories such as von Wright (1971), Menzies and Price (1993), and Price (1991). While these agency approaches try to connect causation to human actions, Woodward's interventionist theory understands interventions in such a broad sense that also phenomena that are not human actions can be interventions. Woodward's theory is based (or, maybe better, inspired) on (or by) the causal nets approach, and especially on (or by) Pearl's (2000) work. To some extent, however, Woodward's theory also exceeds the causal nets approach (cf. Eberhardt and Scheines 2007, p. 981): While the causal nets approach is based on only one basic notion, *viz.* the relation of direct causal connection w.r.t. a given set of variables V, Woodward tries to illuminate causal notions by means of another basic notion: the concept of a possible intervention.[2] The very short (and admittedly intuitively quite satisfying) version of Woodward's interventionist theory amounts to the following: X is a cause of Y w.r.t. a given set of variables V if and only if there is a possible intervention on X such that Y would wiggle when X were wiggled by means of this intervention. This idea can be further specified in different ways such that one can distinguish between three kinds of causal connections: total causation, direct causation, and contributing causation. Woodward defines *total causation* as follows:

> **(TC)** X is a total cause of Y if and only if there is a possible intervention on X that will change Y or the probability distribution of Y. (Woodward 2003, p. 51)

(TC) clearly connects causation to interventions: X is a total cause of Y if and only if Y's value can be changed by intervening on X. Note that this notion of a total cause does not allow for inferring direct causal relationships among variables in a variable set V of interest, i.e., it does not allow us to uncover the causal paths through a set of variables over which X influences Y. Do also note that **(TC)** will fail when it comes to revealing causal connections in case of unfaithful scenarios. If, for example, X causes Y over several different directed causal paths such that these

---

[2]Other researchers, such as Glymour (2004, p. 780), see the theory as providing a philosophically more transparent reformulation of the quite challenging causal nets approach. As we will see later on, Woodward's (2003) theory cannot do that job, since direct causal relations in Woodward's theory coincide with causal arrows within the theory of causal nets often only under very specific conditions.

paths cancel each other exactly, then there will be no intervention on X that will have an influence on Y. Thus, **(TC)** will tell us that X is not related to Y in terms of total causation.

For coping with causal connections in unfaithful scenarios and for uncovering the causal paths over which a variable X may cause another variable Y, Woodward introduces two further causal notions, *viz. direct causation* and *contributing causation*:

> **(M)** A necessary and sufficient condition for X to be a (type-level) direct cause of Y with respect to a variable set **V** is that there be a possible intervention on X that will change Y or the probability distribution of Y when one holds fixed at some value all other variables Z_i in **V**. A necessary and sufficient condition for X to be a (type-level) *contributing cause* of Y with respect to variable set **V** is that (i) there be a directed path from X to Y such that each link in this path is a direct causal relationship [...] and that (ii) there be some intervention on X that will change Y when all other variables in **V** that are not on this path are fixed at some value.[3] (Woodward 2003, p. 59)

Also **(M)** connects causal notions to the concept of a possible intervention. When it is possible to change Y's value by intervening on X, while the values of all variables different from X and Y in a set of variables V are fixed by interventions, then X has to be a direct cause of Y w.r.t. V, and vice versa: When X is a direct cause of Y w.r.t. a given set of variables V, then it must be possible to influence Y's value by intervening on X when the values of all variables different from X and Y in V are held fixed by interventions. The definition of a contributing cause is then based on the notion of direct causation: If a chain of direct causal relationships goes from X to Y—whether this is the case can be tested by appliance of the definition of direct causation given in **(M)**—, one can hold fixed the values of all variables in V that are different from X and Y and that do also not lie on the directed causal path connecting X and Y by interventions, and Y can be wiggled by manipulating X in this context, then X is a contributing cause of Y w.r.t. V. And, again, vice versa: If X is a contributing cause of Y w.r.t. V, then it is possible to hold the values of all off-path variables fixed, while it is possible to influence Y's value by wiggling X. Note that in **(M)** (and also in **(TC)**) Woodward on the one hand speaks of changing Y's value, but on the other hand also of changing Y's probability distribution. This indicates that the theory is intended to allow for deterministic as well as for probabilistic versions of causation. This will become important in Sect. 5.2.2 when I will suggest a certain reconstruction of Woodward's theory.

Do also note the advantages of **(M)** over **(TC)**: Given a variable set V of interest, **(M)**, contrary to **(TC)**, tells us over which causal paths variables in V are causally related. In addition, **(M)** can handle some unfaithful scenarios, at least under the proviso that canceling paths go through variables in V. If X, for example, causes Y directly and over the causal path $X \longrightarrow Z \longrightarrow Y$ in V and influences transported over both paths cancel each other, then X is not a total cause of Y. However, **(M)** tells us that X is a direct (as well as a contributing) cause of Y, because Y's value can be influenced by wiggling X when Z's value is fixed by an intervention.

---

[3]The parts 'when one holds fixed at some value all other variables Z_i in **V**' and 'when all other variables in **V** that are not on this path are fixed at some value' of **(M)** are called fixability conditions.

As we saw, both **(TC)** and **(M)** explicate causation in terms of possible interventions. So for making sense of causation in Woodward's (2003) framework, we also need to have some understanding of the notion of an intervention. To define 'intervention', Woodward first requires the following notion of an intervention variable:

> [...] I is an intervention variable for X with respect to Y if and only if I meets the following conditions: **(IV)** I1. I causes X. I2. I acts as a switch for all the other variables that cause X. That is, certain values of I are such that when I attains one of those values, X ceases to depend on the values of other variables that cause X and instead depends only on the value taken by I. I3. Any directed path from I to Y goes through X. [...] I4. I is (statistically) independent of any variable Z that causes Y and that is on a directed path that does not go through X. (Woodward 2003, p. 98)

Now Woodward proceeds by telling us what interventions, as they appear in **(TC)** and **(M)**, are. $I = i$ is an intervention on X w.r.t. Y if and only if I's taking value i causes (or "forces") X to take a certain value x (cf. Woodward 2003, p. 98).

The main idea behind **(IV)** is to have a notion of an intervention at hand designed in such a way that appliance of **(TC)** and **(M)** guarantees to pick out all and only the corresponding causal relations featured by the world's causal structure. For this purpose, an intervention variable I for X has to be a cause of X (I1), I has to have full control over X when I is on (I2), I must not be allowed to cause Y over a directed path that is not going through X (I3), and I has to be probabilistically independent of any cause of Y causing X over a causal path that does not go through X (I4). For a more detailed discussion and motivation of these conditions see Woodward (2003, sec. 3.1).

Before I will go on and discuss what it means for an intervention to be possible, note two things. First, that Woodward (2003) only provides definitions for the three-place notions of intervention and intervention variable: I is an intervention/intervention variable on/for X w.r.t. Y. (From now on we shall call X the intervention's *target variable* and Y the *test variable*.) Woodward explicitly claims that "there is no such thing as an intervention simpliciter" (Woodward 2003, p. 103). However, do also note that Woodward in his own definitions **(TC)** of total causation and **(M)** of direct and contributing causation only speaks of interventions on X (or Z_i) and not of interventions on X (or Z_i) w.r.t. a test variable. I will follow the interpretation of interventions in such passages as interventions on X (or Z_i) w.r.t. the test variable Y suggested in Gebharter and Schurz (2014).

Second, note that it is still controversial whether Woodward's (2003) interventionist theory of causation contains a vicious circle. Such a circle may arise because Woodward explicates causation in terms of interventions and, vice versa, interventions in terms of causal connections (cf. Strevens 2007). The reaction of Woodward (2008b) (see also Woodward 2009, 2011a) to this objection is that the theory does, indeed, contain a circle, but not a vicious one. The circle is not a vicious one because the questions the theory should allow to answer, *viz.* whether X is a total, direct, or contributing cause of Y relative to a given variable set V, do not require knowledge about whether X is a total, direct, or contributing cause of Y. They only require different causal knowledge, especially knowledge about possibilities of how one can intervene on X and the variables in V different from X and Y. In this chapter, however, I want to bracket the question whether Woodward's response to Strevens is satisfying or not—the results I want to present in this chapter are independent of how this question may be answered.

A problem closely related to the circularity discussion is the question whether Woodward's theory involves an infinite regress: X causes Y only if there is an intervention variable I_X for X w.r.t. Y by whose means Y can be influenced. This presupposes that I_X is a cause of X (intervention condition I1). But then, there has to be an intervention variable I_{I_X} for I_X w.r.t. X by whose means X can be influenced, and so on ad infinitum. I will also bracket this problem.

Let me now come to the notion of a possible intervention. In (TC) and (M) Woodward speaks of possible interventions, while in (IV) he explicates only the notion of an intervention. So he owes us an explanation what it means for an intervention on X w.r.t. Y to be "possible". Woodward (2003, sec. 3.5) uses the following example to demonstrate that possible interventions should not be understood as physically (or nomologically) possible interventions: Assume we are interested in the causal relation between X and Y w.r.t. variable set $V = \{X, Y\}$, where X is a variable standing for the position of the moon, while Y is a variable describing the motion of the tides. Now we want to know whether the position of the moon X is a total, direct, or contributing cause of the motion of the tides Y w.r.t. V. (TC) and (M) tell us that X is a total, direct, or contributing cause of Y w.r.t. V only if there is a possible intervention $I = i$ on X w.r.t. Y that would at least have some probabilistic influence on Y. If this intervention is required to be physically possible, then, of course, there is no such intervention $I = i$. This is the case because changing the moon's position would require an enormous amount of energy. But by the equivalence of mass and energy, every event that may count as an intervention on the position of the moon would also have a direct causal influence on the motion of the tides, what would violate the third Woodwardian intervention condition I3. From this observation Woodward concludes that interventions should not have to be physically possible. They only have to be logically (or conceptually) possible. So what we have to do to check whether X is a total, direct, or contributing cause of Y in V is the following: We have, first, to assume that there is a merely logically possible intervention on X w.r.t. Y, and we have, second, to assess whether this intervention would have an influence on Y. So we have to assess the truth of the following counterfactual: If X were to be manipulated (by such a merely logically possible intervention), then Y's value would change.

To make sense of causal claims (and this is Woodward's 2003 ultimate goal), however, we have to have some basis for assessing the truth of such counterfactuals (cf. Woodward 2003, p. 131, Woodward 2011a, sec. 11). In case of the example discussed above, this basis would be provided by Newtonian gravitational theory and mechanics: The theory tells us that changing the position of the moon (by some merely conceptually possible intervention) would lead to a change in the motion of the tides (when everything else would be fixed by additional and maybe, again, only logically possible interventions). So Newtonian gravitational theory and mechanics tell us that the counterfactual 'if X were to be manipulated, then Y's value would change' is true, and thus, we can infer that the position of the moon X is a cause of the motion of the tides Y w.r.t. $V = \{X, Y\}$.

Though Woodward's (2003) solution seems to work quite well, I want to argue that it actually does not. Assume we want to know whether the motion of the tides is causally relevant to the position of the moon. To establish a causal relationship between the two variables, it suffices, according to Woodward, to have some basis for assessing the truth of the counterfactual claim 'if Y were to be manipulated by means of a (maybe merely logically possible) intervention $I = i$, then X's value would change'. According to Woodward, this can be done by Newtonian gravitational theory and mechanics: The theory tells us that if a change in the motion of the tides were to occur (where everything else were fixed), then a change in the position of the moon would have to occur. So, according to Woodward's idea, it would turn out that the motion of the tides Y is also a cause of the position of the moon X in V, what is obviously false. What one would need here to get things right is some kind of a causally laden version of Newtonian gravitational theory and mechanics. But we do not have such a version; to achieve such a version is rather the goal of applying Woodward's interventionist theory. The conclusion of these observations is that there simply is no basis for determining the effects of merely logically (or conceptually) possible interventions that does not already presuppose causal knowledge. The constraint that possible interventions only have to be logically possible Woodward suggests is too weak.

So in which sense do interventions have to be possible? My claim is that for an intervention to be reasonably called possible it has at least to be physically (or nomologically) possible. Considering a purely logically possible intervention's effects in one's logical imagination is not enough; there has to be, for example, a physically possible process of some kind such that if this process were set in motion, then it would produce the realization of a certain value of the intervention's target variable. Or, there has to be a force of nature exerted by the intervention on its target variable which forces the latter to take a specific value. There might be more ways for an intervention to be physically possible.

As we will see in Sect. 5.3.1, the required characterization of possible interventions as physically possible leads to unattractive problems with Woodward's (2003) interventionist theory. However, it is indispensable for making sense of causal claims. One could even go one step further and argue that interventions should be possible in the sense that they are at least sometimes instantiated in the actual world. For the rest of this chapter, however, I will stick to interventions as physically possible interventions. This interpretation is much closer to Woodward's original ideas behind possible interventions.

5.2.2 Reconstructing Woodward's Interventionist Theory of Causation

In this subsection I will provide reconstructions of Woodward's (2003) notion of an intervention variable and of his three causal notions: total causation, direct causation, and contributing causation. Note that Woodward's theory uses terms such as 'direct causation' and 'causation' which are also used in the theory of causal nets. It is, however, not clear that every direct cause in the sense of Woodward turns out to be a direct cause in the sense of the theory of causal nets (and vice versa). Thus, we have to distinguish the two theories' terminologies. From now on and throughout the whole chapter the terms 'total causation', 'direct causation', and 'contributing causation' will be reserved for the diverse causal concepts explicated in Woodward's theory. The term 'causation*' (with an asterisk) will be used for all three of Woodward's causal notions. All other causal terms (also used throughout earlier chapters), such as 'cause', 'common cause', 'effect', 'direct effect', 'intermediate cause', etc. refer to the graph theoretical notions.

Before I can give reconstructions of Woodward's (2003) three causal notions, I have to introduce two concepts required for these reconstructions: the notion of a causal dependence model and the notion of a causal system's intervention expansion. The first notion will allow us to connect causal relations not only to counterfactuals about what would happen under interventions that are at least sometimes instantiated in our world, but also to counterfactuals about what would happen under interventions that are physically possible but never instantiated in the actual world. This is required to keep the reconstruction as close to Woodward's original theory as possible. The second notion tells us which conditions must be satisfied when choosing intervention variables for accounting for causal connections. Intervention expansions are not explicitly considered in Woodward (2003). But, as we will see, such a notion is required, because adding arbitrarily chosen intervention variables to the variables in whose causal connections we are interested may alter or destroy these causal relations.

In his definitions **(TC)** and **(M)**, Woodward (2003) is clearly interested in the causal relationships among the elements of certain sets of variables V. In causal nets terminology this means that Woodward is interested in the causal structure $\langle V, E \rangle$ underlying V. This causal structure, however, can only be identified by its empirical footprints. Till now we represented causation's empirical footprints by empirical models $\langle V_e, P_e \rangle$. But since Woodward's theory is so closely connected to counterfactuals about what would happen under (physically) possible interventions, an empirical model just featuring the actual world's probability distribution P_e is not sufficient for a reconstruction of Woodward's theory. To see this, assume X to be a cause of Y w.r.t. a variable set V. Assume further that interventions on X w.r.t. Y never occur in the actual world. This is possible for Woodward. His theory can still account for X's being a cause of Y w.r.t. V. Everything needed for this is that in some (physically) possible worlds some interventions on X w.r.t. Y occur and have an influence on Y. If our reconstruction of Woodward's theory only refers to the probability distribution of the actual world, then, because there are no interventions on X w.r.t. Y that influence Y in the actual world, our reconstruction would tell us, contrary to Woodward's original theory, that X is not a cause of Y w.r.t. V. Thus, to be able to account for causation in a way similar to Woodward, we have to choose a type of causal model that allows us to capture dependencies between variables and (physically) possible interventions that never occur in our world.

All the counterfactual information required here can be provided by a causal dependence model, whose notion is based on the concept of a dependence model DEP rather than on the notion of an empirical model $\langle V_e, P_e \rangle$. Dependence models abstract from specific probabilistic details and, instead, provide only information about certain dependence and independence relations shared by all the probability distributions of interest[4]:

Definition 5.1 (dependence model) DEP is a dependence model over variable set V if and only if DEP is a set of (in)dependence relations $(IN)DEP(X,Y|Z)$ among variables (or sets of variables) in V.

Given a set of specified probability distributions P over a set of variables V and a dependence model DEP over V, '$DEP(X,Y|Z)$' stands short for 'X and Y are probabilistically dependent conditional on Z in every one of these distributions P', and '$INDEP(X,Y|Z)$' stands short for 'X and Y are probabilistically independent conditional on Z in every one of these distributions P'. Dependence models DEP over V from now on shall represent physically possible probability distributions over V featuring the probabilistic (in)dependencies in DEP.

A dependence model DEP over a set of variables V together with a causal structure $\langle V, E \rangle$ becomes a *causal dependence model* $\langle V, E, DEP \rangle$. Causal dependence models $\langle V, E, DEP \rangle$ from now on are used to describe physically possible worlds with causal structure $\langle V, E \rangle$ whose probability distributions' (in)dependencies are described by DEP.[5]

Let us now take a closer look at how exactly causal dependence models can avoid the problem outlined above: Suppose we are interested in $V = \{X, Y\}$, but no intervention $I_X = ON$ on X w.r.t. Y ever occurs in the actual world. However, let us assume we have a dependence model DEP over V. When adding intervention variables I_X for X to V, there are basically two kinds of expansions $\langle V', E', DEP' \rangle$ of the causal dependence model $\langle V, E, DEP \rangle$ in which we may be interested: (i) expansions $\langle V', E', DEP' \rangle$ in which $DEP'(Y, I_X = ON)$ holds, and (ii) expansions in which $INDEP'(Y, I_X = ON)$ holds. Expansions of type (ii) describe worlds like ours in which $I_X = ON$ is never instantiated or does have no influence on Y, while expansions of type (i) describe worlds in which $I_X = ON$ is sometimes instantiated and in whose probability distributions some Y-values are correlated with $I_X = ON$. The reconstructions of Woodward's causal notions developed in this subsection will

---

[4]The notion of a dependence model is borrowed from Pearl et al. (1990, p. 509).

[5]I will call physically possible worlds with causal structure $\langle V, E \rangle$ sometimes $\langle V, E \rangle$-physically possible.

have to say something like this: If our causal dependence model can be expanded to a model of kind (i), i.e., if there are physically possible interventions $I_X = ON$ on X w.r.t. Y that have some influence on Y, then X is a cause of Y in V. Otherwise, X does not cause Y w.r.t. V. This solves the above-mentioned problem.

As we have seen above, testing for whether X is a cause of Y requires adding certain intervention variables to one's set of variables of interest V. So for testing causal relationships among variables of a set V we have to expand this set in a certain way. But which expansions are allowed? Are there any restrictions when it comes to adding intervention variables to V? Woodward (2003) does not provide an explicit answer to this question. It seems, however, clear that causal information, i.e., information about which variables are causally relevant for which variables, must be preserved. In other words: The causal information stored by our original system's causal structure $\langle V, E \rangle$ must fit to the expansion's causal structure. This is guaranteed by the following notion of an expansion (taken from Steel 2005, p. 11):

Definition 5.2 (expansion of a causal structure) If $\langle V, E \rangle$ is a causal structure, then $\langle V', E' \rangle$ is an expansion of $\langle V, E \rangle$ if and only if

(a) $V \subseteq V'$, and
(b) for all $X, Y \in V$: If $X \longrightarrow Y$ in $\langle V', E' \rangle$, then $X \longrightarrow Y$ in $\langle V, E \rangle$, and
(c) for all $X, Y \in V$: If there are $Z_1, \ldots, Z_n$ in V' all of which are not in V such that $X \longrightarrow Z_1 \longrightarrow \ldots \longrightarrow Z_n \longrightarrow Y$ in $\langle V', E' \rangle$, then $X \longrightarrow Y$ in $\langle V, E \rangle$, and
(d) no causal path not implied by (c) and (d) is in $\langle V, E \rangle$.

According to Definition 5.2, every causal structure is an expansion of itself. Condition (a) guarantees that the expansion's variable set contains all variables of the original system. Conditions (b)–(d) guarantee that the expansion's causal structure fits the original model's causal structure. (b) tells us that two variables X and Y represented in both systems stand in direct causal relation in the original structure whenever they do so in the expanded structure, and (c) that they do also stand in direct causal connection in the original structure if they are connected via a directed causal path in the expansion such that all intermediate causes lying on this path are not elements of the original structure's set of vertices. (d) finally states that the causal relationships implied by the expansion together with (b) and (c) are all the causal relationships featured by the original system's causal structure. Hence, there are no causal relations in $\langle V, E \rangle$ that are not in some way transferred to $\langle V', E' \rangle$ when expanding $\langle V, E \rangle$.

Note that the given notion of an expansion only preserves causal information about which variables cause which variables. (This is exactly what Woodward's theory requires.) When one goes from the expansion to the original structure, one can marginalize out intermediate causes, common causes, or common effects. When marginalizing out an intermediate cause Z lying on a directed path from X to Y in the expansion ($X \longrightarrow\longrightarrow Z \longrightarrow\longrightarrow Y$), X will still be a cause of Y in the original structure ($X \longrightarrow\longrightarrow Y$). However, if one marginalizes out a common cause or common effect Z of X and Y, then these pieces of causal information will not be preserved.[6]

Based on Definition 5.2, we now can define the notion of an expansion of a causal dependence model:

Definition 5.3 (expansion of a causal dependence model) If $S = \langle V, E, DEP \rangle$ is a causal dependence model, then $S' = \langle V', E', DEP' \rangle$ is an expansion of S if and only if

(a) $\langle V', E' \rangle$ is an expansion of $\langle V, E \rangle$, and
(b) $DEP' \uparrow V = DEP$.[7]

This latter notion of an expansion does not only preserve information about which variables cause which variables, but also ensures that the dependence and independence relations featured by the original and by the expanded model fit together. Note that expanding a causal dependence model in the vein of Definition 5.3 presupposes that the model's context $C = c$ is not changed when expanding, because changing (or lifting) the context may lead to violations of condition (b) in Definition 5.3.

When it comes to adding intervention variables to one's system of interest, we will (following Spirtes et al. 2000, sec. 3.7.2) refer to the original system as the *unmanipulated system*, and to the expanded system containing the required intervention variables as the *manipulated system*. When going from the unmanipulated to the manipulated system, we will typically change the system's context. The unmanipulated system's context $C = c$ contains intervention variables realizing their OFF-values for all individuals in the domain, while these intervention variables' values are allowed to vary in the manipulated system.[8] This representation has some advantages over the typical "arrow breaking" terminology (cf. Pearl 2000, sec. 1.3.1) in the literature. Its main advantage is that the unmanipulated model's probability distribution as well as the probabilities of value instantiations of variables in V under all possible combinations of values of intervention variables can be computed in the manipulated model while the system's causal structure is preserved.

Based on the considerations above we now define two notions of an intervention expansion ('i-expansion' for short):

Definition 5.4 (strong i-expansion) If $S = \langle V, E, DEP \rangle$ is a causal dependence model, then $S' = \langle V', E', DEP' \rangle$ is a strong i-expansion of S w.r.t. test variable Y if and only if

---

[6]For a notion of an expansion that preserves also information about common causes, see Definition 6.2 in Sect. 6.4 (cf. Gebharter 2014, p. 147).

[7]$DEP' \uparrow V$ is the restriction of DEP' to V. The restriction of DEP' to V contains all and only the (in)dependencies in DEP' among variables (or sets of variables) in V.

[8]Another way to go would be to allow for intervention variables which take their ON-values for some individuals in the domain in the unmanipulated model. But in this chapter I follow Spirtes et al. (2000, sec. 3.7.2) in assuming that the intervention variables are OFF for all individuals in the unmanipulated model, which is the standard way to go in the literature.

(a) $\langle V', E' \rangle$ is an expansion of $\langle V, E \rangle$, and

(b) $V' \backslash V$ contains for every $Z \in V$ different from Y an intervention variable I_Z for Z w.r.t. Y (and nothing else), and

(c) for all $Z_i, Z_j \in V$ and all $M \subseteq V \backslash \{Z_i, Z_j\}$: $(IN)DEP(Z_i, Z_j | M)$ if and only if $(IN)DEP'(Z_i, Z_j | M, I = OFF)$, where I is the set of all intervention variables in $V' \backslash V$, and

(d) for every $\langle V, E \rangle$-physically possible probability distribution P over V represented by DEP and every $z \in val(Z)$ of every $Z \in V \backslash \{Y\}$ there is an ON-value i_Z of the corresponding intervention variable I_Z for Z w.r.t. Y such that $DEP'(z, I_Z = i_Z)$ and $P(z | I_Z = i_Z) = 1$.

The notion of a strong i-expansion is suitable for accounting for direct and contributing causation by means of interventions. Condition (a) requires that adding intervention variables does not alter the causal information provided by the unmanipulated system. Condition (b) tells us that the manipulated system contains an intervention variable for every variable in the original system different from the test variable Y w.r.t. this test variable Y, meaning that we can fix the value of any variable different from Y by interventions in the i-expansion. This is required by the fixability conditions for direct and contributing causation in **(M)**. Condition (c) tells us how the manipulated and the unmanipulated system's associated dependence models fit together: The (in)dependencies among variables in V in the unmanipulated system should be the same as in the manipulated system when all intervention variables I in $V' \backslash V$ have taken their OFF-values. So strong i-expansions are not expansions in the sense of Definition 5.3; they allow at least for a partial lifting of the unmanipulated system's context $C = c$. Due to condition (d), every variable Z different from the test variable Y can be forced to take every one of its possible values by an intervention variable I_Z for Z w.r.t. Y. This is required to guarantee that the non-target and the non-test variables $Z \in V$ can be fixed to arbitrary value combinations. Without this possibility, it would not be assured that non-dependence of Y and an intervention $I_X = ON$ on X w.r.t. Y indicates that X is not a direct (or contributing) cause of Y when all other intervention variables (or all intervention variables for off-path variables) have taken their ON-values. Without this requirement it could still be that I_X's ON-values are only correlated with X-values which are not correlated with Y-values, or that Y's X-alternative parents can only be forced to take such values which cancel X's direct (or contributing) causal influence on Y.

Definition 5.5 (weak i-expansion) If $S = \langle V, E, DEP \rangle$ is a causal dependence model, then $S' = \langle V', E', DEP' \rangle$ is a weak i-expansion of S for target variable X w.r.t. the test variable Y if and only if

(a) $\langle V', E' \rangle$ is an expansion of $\langle V, E \rangle$, and

(b) $V' \backslash V$ contains an intervention variable I_X for X w.r.t. Y (and nothing else), and

(c) for all $Z_i, Z_j \in V$ and all $M \subseteq V \backslash \{Z_i, Z_j\}$: $(IN)DEP(Z_i, Z_j | M)$ if and only if $(IN)DEP'(Z_i, Z_j | M, I_X = OFF)$, and

(d) for every $\langle V, E \rangle$-physically possible probability distribution P over V represented by DEP and every $x \in val(X)$ there is an ON-value i_X of the corresponding intervention variable I_X for X w.r.t. Y such that $DEP'(x, I_X = i_X)$ and $P(x | I_X = i_X) = 1$.

The notion of a weak i-expansion is suitable for accounting for total causation by means of interventions. Condition (a) does the same work as condition (a) in Definition 5.4. Condition (b) is weaker than condition (b) in Definition 5.4. It only requires one intervention variable I_X for the target variable X w.r.t. test variable Y. This suffices for testing for total causation, since testing for total causation between X and Y does not require that other variables in V are fixed by interventions. Thus, also condition (c) can be weaker. It just requires that the (in)dependencies among variables in V in the unmanipulated system have to be the same as in the manipulated system when $I_X = OFF$. Thus, also weak i-expansions allow for partially lifting the unmanipulated system's context and, hence, are not expansions in the sense of Definition 5.3. Last but not least, condition (d) can also be a little bit weaker than its counterpart in Definition 5.4. It only requires full control of the target variable X.

I will proceed by giving a reconstruction of Woodward's (2003) definition of an intervention variable (**IV**). Next I will show how this reconstruction together with the above introduced versions of an i-expansion can be used to reconstruct Woodward's notions of total, direct, and contributing causation. Here comes the reconstruction of (**IV**):

Definition 5.6 (intervention variable) If $S = \langle V, E, DEP \rangle$ is a causal dependence model with $I_X, X, Y \in V$, then I_X is an intervention variable for X w.r.t. Y in S if and only if

(a) $I_X \longrightarrow X$ in S, and
(b) for every $\langle V, E \rangle$-physically possible probability distribution P over V represented by DEP and every $i_X \in ON$ there is an $x \in val(X)$ such that $DEP_P(x, i_X)$ and $P(x | i_X) = 1$, and
(c) all paths $I_X \longrightarrow\!\!\!\longrightarrow Y$ in S go through X, and
(d) $INDEP(I_X, Z)$ holds for all Z that cause Y over a directed path that does not go through X in any expansion $S' = \langle V', E', DEP' \rangle$ of S.

Condition (a) mirrors Woodward's (2003) condition I1: For I_X to be an intervention variable for X w.r.t. Y, I_X has to be a cause of X.[9] By 'cause' in I1 Woodward means a contributing cause (cf. Woodward 2003, p. 98), which can be explicated by conditions (a) and (b). Condition (b) tells us that an intervention variable's values must be distinguishable in two different classes: ON- and OFF-values. An intervention variable I_X's ON-values are those values that force X to take a certain value x when attained. I_X's OFF-values are all values in $val(I_X)$ that are independent

---

[9]The assumption that I_X is a causal parent of X in condition (a) is a harmless simplification of I1. If I_X would be an indirect cause of X in a dependence model, then there is always also another dependence model in which I_X is a causal parent of X.

of X. '$I_X = ON$' and '$I_X = OFF$' stand short for 'I_X has taken one of its ON-values' and 'I_X has taken one of its OFF-values', respectively. Note that $I_X = ON$ will screen X off from all its causes not lying on a path $I_X \longrightarrow\!\!\!\longrightarrow X$ in S and, thus, implies Woodward's condition I2. Condition (c) requires, as also Woodward's original condition I3 does, that all causal paths from I_X to Y in S go through X. Condition (d) mirrors Woodward's condition I4: To infer that Y has been influenced by a change in X brought about by $I_X = ON$ one has to exclude that I_X and Y are correlated over common cause paths or somehow else. Note that (d) requires not only independence of variables in V, but independence of variables Z in any expansion of S (cf. Woodward 2008b, p. 202). So what condition (d) is intended to do is to assure (together with condition (c)) that all changes in Y brought about by $I_X = ON$ are produced by a directed causal path from I_X to Y that goes through X.

With help of the notion of a causal dependence model, the concept of an i-expansion, and the above reconstruction of **(IV)**, we are now able to give straight-forward reconstructions of Woodward's (2003) three kinds of causal relationships. I begin with proposing the following reconstruction of Woodward's notion of total causation (**TC**):

Definition 5.7 (total causation) If $S = \langle V, E, DEP \rangle$ is a causal dependence model with $X, Y \in V$, then X is a total cause of Y in S if and only if $DEP'(Y, I_X = ON)$ holds in some weak i-expansions $S' = \langle V', E', DEP' \rangle$ of S for X w.r.t. Y, where I_X is an intervention variable for X w.r.t. Y in S'.

According to Definition 5.7, for establishing a total causal relationships between X and Y in a causal dependence model $S = \langle V, E, DEP \rangle$ it is required that there is a weak i-expansion of S for X w.r.t. Y that features an intervention variable I_X for X w.r.t. Y such that Y depends on $I_X = ON$. So what we are doing when testing whether X is a total cause of Y in S is the following: We check whether there is a physically possible world whose causal structure fits to $\langle V, E \rangle$ and which features an intervention variable I_X for X w.r.t. Y such that this world's probability distribution P features the (in)dependencies in DEP when $I_X = OFF$ and such that $P(y, I_X = ON) \neq P(y)$ holds for some Y-values y in this world's distribution P. If there is such a world, then X is a total cause of Y in S. If there is no such world, then X is not a total cause of Y in S.

Note that X's being a total cause of Y and X's being a cause of Y (in the graph theoretical sense, i.e., $X \longrightarrow\!\!\!\longrightarrow Y$) can be expected to coincide in case the causal dependence model $\langle V, E, DEP \rangle$ satisfies the causal Markov condition (CMC) and the causal faithfulness condition (CFC). In non-faithful scenarios, Woodward's notion of total causation and the graph theoretical notion of causation can be expected to deviate from each other (recall the discussion of total causation in Sect. 5.2.1). This is because there may be causal paths $X \longrightarrow\!\!\!\longrightarrow Y$ in non-faithful causal systems though X does not depend on Y. In such cases, according to Definition 5.7, X will not be a total cause of Y w.r.t. V.

Here comes the reconstruction of Woodward's (2003) notion of direct causation. This reconstruction is inspired by (Gebharter and Schurz 2014, sec. 4):

Definition 5.8 (direct causation) If $S = \langle V, E, DEP \rangle$ is a causal dependence model with $X, Y \in V$, then X is a direct cause of Y in S if and only if $DEP'(Y, I_X = ON | I_Z = ON)$ holds in some strong i-expansions $S' = \langle V', E', DEP' \rangle$ of S w.r.t. Y, where I_X is an intervention variable for X w.r.t. Y in S' and I_Z is the set of all intervention variables for variables $Z \in V$ (different from X and Y) w.r.t. Y in S'.

According to this definition, X is a direct cause of Y in a causal dependence model $S = \langle V, E, DEP \rangle$ whose set of vertices contains X and Y if and only if the system S can be expanded in a certain way. In particular, there has to be a strong i-expansion $S' = \langle V', E', DEP' \rangle$ w.r.t. Y such that V' contains an intervention variable I_X for X w.r.t. Y as well as intervention variables I_Z for all Z in V different from X and Y. Now this expanded system's dependence model DEP' has to feature a dependence between Y and $I_X = ON$ given that all other intervention variables I_Z have taken certain ON-values. The idea is exactly the same as in **(M)**: X is a direct cause of Y in S if and only if $I_X = ON$ has some influence on Y in S' when the values of all variables $Z \in V$ different from X and Y are fixed by interventions $I_Z = ON$.

So to establish a direct causal connection between X and Y in a causal dependence model $S = \langle V, E, DEP \rangle$, one has to check whether there is a physically possible world whose causal structure fits to $\langle V, E \rangle$, whose probability distribution P fits to the probabilistic (in)dependencies in DEP among variables in V when $I_X = OFF$ and $I_Z = OFF$, and which features a probabilistic dependence between some Y-values y and the instantiation $I_X = ON$ of an intervention variable I_X for X w.r.t. Y when the values of all other variables $Z \in V$ are fixed by interventions $I_Z = ON$. The latter means that one has to compare the conditional probabilities $P(y | I_X = ON, I_Z = ON)$ and $P(y | I_Z = ON)$ in this world's distribution P.[10]

Note that the notion of direct causation cannot be expected to coincide with every causal arrow in any causal dependence model. It should coincide with causal arrows $X \longrightarrow Y$ in systems satisfying CMC and the causal minimality condition (Min). In systems not satisfying (Min), X's being a direct cause of Y will deviate from $X \longrightarrow Y$. In such systems there will be an unproductive causal arrow $X \longrightarrow Y$ and, hence, Definition 5.8 will determine X not to be a direct cause of Y.

Here comes the reconstruction of the last one of Woodward's (2003) three notions of causation, i.e., contributing causation:

---

[10]Note that one could also compare the conditional probabilities $P(y | I_X = ON, I_Z = ON)$ and $P(y | I_X = i_X^*, I_Z = ON)$, where i_X^* is either an ON-value different from $I_X = ON$ or an OFF-value of I_X. Zhang and Spirtes' (2008, p. 243) definition of direct causation, for example, which seems to be also inspired by Woodward's (2003), compares the probability of y conditional on two different ON-values of I_X when I_Z is ON. Woodward is not very clear about which pair of probabilities should be compared and there is some room for speculations. However, which pair one compares for assessing direct causal relationships does not have an influence on the results I want to present in this chapter. So I stick with the comparison of $P(y | I_X = ON, I_Z = ON)$ and $P(y | I_Z = ON)$, which seems to me to be the one closest to Woodward's wording in **(M)**.

Definition 5.9 (contributing causation) If $S = \langle V, E, DEP \rangle$ is a causal dependence model with $X, Y \in V$, then X is a contributing cause of Y in S if and only if there is a strong i-expansion $S' = \langle V', E', DEP' \rangle$ of S w.r.t. Y such that

(a) there is a chain π of direct causal relations (in the sense of Definition 5.8) from X to Y in S, and
(b) $DEP'(Y, I_X = ON | I_Z = ON)$, where I_X is an intervention variable for X w.r.t. Y in S' and I_Z is the set of all intervention variables for variables $Z \in V$ (not lying on path π) w.r.t. Y in S'.

Also this reconstruction is very close to Woodward's original notion of contributing causation in **(M)**: For X to be a contributing cause of Y in a system $S = \langle V, E, DEP \rangle$ there first has to be a strong i-expansion of the system featuring a directed causal path π consisting of direct causal relations (in the sense of Definition 5.8) from X to Y. Second, the i-expansion has to contain an intervention variable I_X for X w.r.t. Y and V' has to contain intervention variables I_Z for those Z in V that do not lie on π. Now we can make a similar test as in the case of testing for direct causal dependence: We just have to check whether $I_X = ON$ has some influence on Y when all off-path variables' values are fixed by interventions $I_Z = ON$ in some strong i-expansions. If we find such i-expansions S', then X is a contributing cause of Y in S; if there is no such i-expansion S', then X is not a contributing cause of Y in S.

Note that Woodward's (2003) notion of contributing causation cannot be expected to coincide with the existence of a directed causal path $X \longrightarrow\!\!\longrightarrow Y$ in every causal dependence model satisfying CMC. The notion of contributing causation is intended to pick out only such directed causal paths $X \longrightarrow\!\!\longrightarrow Y$ that are productive (under interventions $I_X = ON$ on X and $I_Z = ON$ on off-path variables). X should not turn out to be a contributing cause of Y in case X causes Y only over unproductive causal chains $X \longrightarrow\!\!\longrightarrow Y$. But how can we distinguish between productive and non-productive directed causal paths π without using interventions? In other words: How can we specify directed causal paths whose existence indicates a contributing causal relationships without already making use of the notion of an intervention? In case of indirect causal paths, there is no straightforward solution to the problem of identifying such a path's productiveness. (For an attempt to specify productiveness for all kinds of causal paths, see Nyberg and Korb 2006, sec. 7).

However, we can identify a subclass of causal dependence models in which the presence of a directed path from X to Y should coincide with contributing causation à la Woodward (2003): If a causal dependence model features a directed path π from X to Y, then there should be a set M of variables that isolates this path, i.e., that blocks all paths between X and Y but π. In such a causal dependence model X can be expected to be a contributing cause of Y in case X and Y are dependent conditional on M. For an example of such a causal system, see Fig. 5.1: In this system the presence of a directed path is expected to coincide with contributing causation. Every directed causal path between X and Y, for example, can be isolated and is assumed to be productive. $M(\pi) = \{Z_3, Z_4\}$ of $\pi : X \longrightarrow Z_1 \longrightarrow Z_2 \longrightarrow Y$

Fig. 5.1 Every directed path π between X and Y can be isolated by conditionalizing on some set M of variables

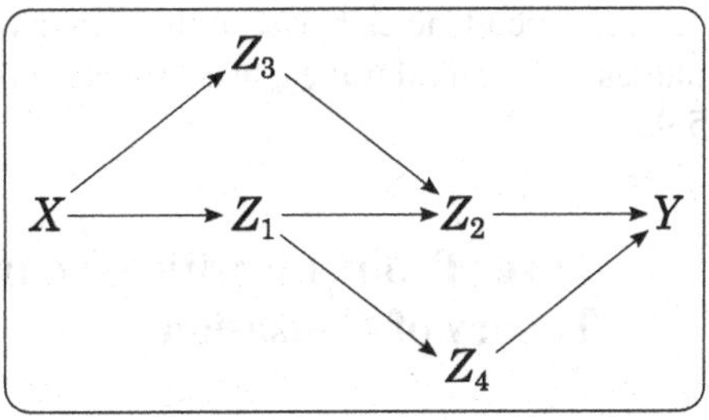

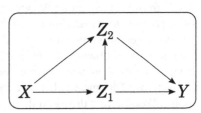

Fig. 5.2 Here not every directed causal path connecting X and Y can be isolated

blocks $\pi^* : X \longrightarrow Z_3 \longrightarrow Z_2 \longrightarrow Y$ and $\pi^{**} : X \longrightarrow Z_1 \longrightarrow Z_4 \longrightarrow Y$, $M(\pi^*) = \{Z_1, Z_4\}$ blocks π and π^{**}, and $M(\pi^{**}) = \{Z_2, Z_3\}$ blocks π and π^*. For an example of a system not satisfying the requirements above, see Fig. 5.2: To isolate $X \longrightarrow Z_1 \longrightarrow Y$, one has to conditionalize on Z_2. But while conditionalizing on Z_2 blocks path $X \longrightarrow Z_2 \longrightarrow Y$, it activates another path between X and Y, *viz.* path $X \longrightarrow Z_2 \longleftarrow Z_1 \longrightarrow Y$.

Before I will go on to the next section and highlight some problems for Woodward's (2003) interventionist theory of causation, let me say some words on a possible objection to the reconstruction of Woodward's three causal notions suggested in this section. One may argue that the reconstructions of the definitions of total, direct, and contributing causation suggested are circular. A causal* relationship between X and Y is defined relative to a causal dependence model $S = \langle V, E, DEP \rangle$ and requires reference to an i-expansion $S' = \langle V', E', DEP' \rangle$ of this model. But $S = \langle V, E, DEP \rangle$ (and, in some sense, also $S' = \langle V', E', DEP' \rangle$) already contains the causal structure $\langle V, E \rangle$. So X is a cause* of Y w.r.t. S if and only if X causes Y in S as well as in S' and certain additional requirements are fulfilled. Thus, to know that X is a cause* of Y w.r.t. S we already have to know that X causes Y in S.

I think that such an argumentation would be based on a misunderstanding. First of all, recall that we distinguished between Woodward's (2003) notions of causation* and the different notions of causation within the theory of causal nets. It cannot be taken for granted that Woodward's notions coincide with causal connections within the theory of causal nets. Second, and more importantly, learning that X is a cause* of Y w.r.t. S does not require more causal information than in Woodward's original theory (introduced in Sect. 5.2). We do not have to know the causal structure of the system of interest S or of S's i-expansion S'. We just have to know that S' features the right intervention variables with the right dependence properties. When we observe certain dependencies between these intervention variables and the test variables of interest, we can infer the causal structure of S by means of Definitions 5.7, 5.8, and 5.9.

5.3 Three Problems with Woodward's Interventionist Theory of Causation

Recall from Sect. 5.2.2 that Woodward's (2003) three notions of causal connection should be able to account for certain kinds of causal relationships in causal systems satisfying certain combinations of CMC, Min, and CFC. What I will do in this section is the following: I will present three kinds of causal scenarios in which at least one of Woodward's three definitions does not single out the right causal connections. This means that there are several causal scenarios such that if one applies Woodward's theory to these scenarios, the theory will give us results that deviate from the assumed causal structures. Thus, the theory must be modified to correctly account for all three intended kinds of causal relationships. For each problem I will suggest and discuss several such modifications.

Note that in the remainder of this section I will implicitly assume CMC and faithfulness, CMC and Min, and CMC and the possibility of isolating every directed causal path when speaking of total, direct, and contributing causation, respectively. These combinations of pre-requirements will become explicit again in Sect. 5.4.

5.3.1 Problem (i): Causes One Cannot Intervene on

The first problem is quite well known. It arises due to the bi-conditional in the definitions of total, direct, and contributing causation (Definitions 5.7, 5.8, and 5.9, respectively). For a variable X to be a cause* of another variable Y in a given system S it is required, according to these definitions, that there is an i-expansion of S containing an intervention variable for X w.r.t. Y such that Y can be influenced by this latter intervention variable's taking certain ON-values. The problem is that there may be causal systems for which there are no expansions containing the required intervention variables. Take the scenario discussed in Sect. 5.2.1 as an example: We know that the position of the moon X is a cause of the motion of the tides Y in causal system S with variable set $V = \{X, Y\}$. But Woodward's (2003) theory would lead to the false result that it is not. (See the discussion about possible interventions in Sect. 5.2.) Even if one is not satisfied by the argumentation in Sect. 5.2 that possible interventions should be understood as "physically possible" and insists on interpreting them as logically (or conceptually) possible, like Woodward does, the problem remains: There are still causes one cannot possibly intervene on, not even in the weak sense of logical or conceptual possibility. Examples of such causes are species, race, sex, etc. (cf. Glymour 2004, pp. 789f).

In my view, the best solution to the problem should not consist in tinkering with the notion of a possible intervention, but rather in modifying the principles which determine the connection of causation to manipulation and control. It is quite reasonable to assume that control of Y by manipulating X indicates a causal connection between X and Y, but why should any such causal relationship between X and Y require the possibility of affecting Y by means of manipulating X? For some variables there simply are no (physically or even conceptually) possible interventions. According to this consideration, one could simply drop the only-if part in Definitions 5.7, 5.8, and 5.9:

Modification 5.1 *Drop the only-if part of Definitions* 5.7, 5.8, *and* 5.9.

Though Modification 5.1 does avoid the problem described above and allows for appliance of the interventionist theory to systems featuring variables such as species, race, sex, etc., it does not lead to any restriction when it comes to causal inference. If the system of interest can be expanded in such a way that the expansion contains the right intervention variables, then the same causal conclusions can be drawn—the same causal relationships can be inferred.

What is not possible anymore are inferences to negative causal facts. We cannot, for example, infer that X is not a cause* of Y in a given system S in case there is no i-expansion containing an intervention variable for X w.r.t. Y by whose means Y's probability distribution could be changed. However, this is not such a big step back, at least not when we are mainly interested in practical research. Even without Modification 5.1 we would somehow have to guarantee that there is no i-expansion that contains the right intervention variables for drawing the conclusion that X does not cause* Y. What we would have to establish, thus, is a negative existential.

However, the main shortcoming of Modification 5.1 is that it gives us only sufficient conditions for total, direct, and contributing causation. One could say that this violates the main aim of Woodward's (2003) theory, *viz.* explicating the meaning of causation. Instead of accepting Modification 5.1, one could also replace Woodward's explicit definitions of total, direct, and contributing causation by partial definitions whose if-conditions presuppose suitable intervention expansions. (For a similar suggestion, see Baumgartner 2010, sec. 4.) This move still allows to explicate the meaning of the three causal notions for systems for which the required intervention expansions exist. In other systems, one would have to account for causation on the basis of non-interventionist terms, e.g., by means of the causal Markov condition (CMC) and the causal minimality condition (Min).

So one could also accept the following modification:

Modification 5.2 *Replace the explicit Definitions* 5.7, 5.8, *and* 5.9 *by the following partial definitions, respectively:*

Total causation: *If there exist weak i-expansions* $S' = \langle V', E', DEP' \rangle$ *of* $S = \langle V, E, DEP \rangle$ *for* X *w.r.t.* Y, *then:* X *is a total cause of* Y *in* S *if and only if* $DEP'(Y, I_X = ON)$, *where* I_X *is an intervention variable for* X *w.r.t.* Y *in* S'.

Direct causation: *If there exist strong i-expansions* $S' = \langle V', E', DEP' \rangle$ *of* $S = \langle V, E, DEP \rangle$ *w.r.t. Y, then* $X \in V$ *is a direct cause of Y in S if and only if* $DEP'(Y, I_X = ON | I_Z = ON)$, *where* I_X *is an intervention variable for X w.r.t. Y in* S' *and* I_Z *is the set of all intervention variables in* S' *different from* I_X.

Contributing causation: *If there exists a strong i-expansion* $S' = \langle V', E', DEP' \rangle$ *of* $S = \langle V, E, DEP \rangle$ *w.r.t. Y, then* $X \in V$ *is a contributing cause of Y in S if and only if*

(a) *there is a chain* π *of direct causal relations (as defined above) from X to Y in S, and*

(b) $DEP'(Y, I_X = ON | I_Z = ON)$, *where* I_X *is an intervention variable for X w.r.t. Y in* S' *and* I_Z *is the set of all intervention variables for variables* $Z \in V$ *(not lying on path* π*) w.r.t. Y in* S'.

Modification 5.2 does not lead to false results when applied to systems containing variables one cannot manipulate. But it does also not give us any information about causation in such systems—our modified interventionist theory keeps silent about causation in such systems simply because the if-conditions of the partial definitions of total, direct, and contributing causation would not be satisfied in that case.

Modification 5.2 keeps also silent about causation in systems featuring variables one cannot fully control by means of intervention variables as defined in Definition 5.6. Such systems are systems featuring variables which are only manipulable in a weak stochastic sense. An example is the decay of uranium which can only be probabilistically influenced (cf. Reutlinger 2012, p. 789). Another example is given by Campbell (2007, sec. 3). Campbell argues that certain psychological systems contain variables one cannot manipulate by intervention variables that satisfy the second intervention condition (b) in Definition 5.6. Certain actions are, for example, definitely caused by our intentions to carry them out. However, there is no intervention variable that could fully determine our intentions. Our intentions are always partially influenced by rational causes such as our beliefs, our priorities, etc. The influence of these rational causes on our intentions cannot be broken, but only in- or decreased by such external interventions.

To make Woodward's (2003) interventionist theory of causation applicable to systems containing variables which can only be influenced in a probabilistic way, we have to allow for so-called *stochastic interventions* (cf. Korb et al. 2004, sec. 5). We can do so by replacing condition (b) of Definition 5.6 with a condition that does not require an intervention variable to force its target variable to take a certain value when *ON* anymore:

Modification 5.3 *Replace condition (b) in Definition 5.6 by: (b*) for every* $i_X \in ON$ *and every* $\langle V, E \rangle$*-physically possible probability distribution P over V represented by DEP there is an* $x \in val(X)$ *such that* $P(x | i_X) \neq P(x)$.

This latest modification does not only allow for appliance of the interventionist theory to causal systems containing variables one cannot deterministically intervene on, but does, unfortunately, also lead to a new problem with the notion of direct causation (but not with total or contributing causation).

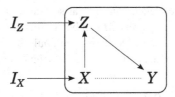

Fig. 5.3 For X to be a direct cause of Y in $S = \langle V, E, DEP \rangle$ it suffices that there is a strong *i*-expansion of S for Y featuring an intervention variable I_X for X w.r.t. Y and an intervention variable I_Z for Z such that Y depends on $I_X = ON$ conditional on $I_Z = ON$. When accepting Modification 5.3, $I_Z = ON$ may be a stochastic intervention that does not freeze Z's value and, thus, does not block influence of I_X on Y over path $I_X \longrightarrow X \longrightarrow Z \longrightarrow Y$. Hence, the modified theory may lead to the false result that X is a direct cause of Y in S

Assume we are interested in a causal system $S = \langle V, E, DEP \rangle$ whose variable set V is $\{X, Y, Z\}$ and whose causal structure is $X \longrightarrow Z \longrightarrow Y$. (See Fig. 5.3 for a graphical illustration.) In this causal structure, X is not a causal parent of Y. Now assume that we want to apply our interventionist theory to the system. We ask the theory what it can tell us about the causal relationship between X and Y. In particular, we ask the theory whether X is a direct cause of Y in S. Definition 5.8 tells us that X is a direct cause of Y if and only if there is a strong *i*-expansion $S' = \langle V', E', DEP' \rangle$ of S for Y such that V' contains an intervention variable I_X for X w.r.t. Y as well as an intervention variable I_Z for Z such that $P(y|I_X = ON, I_Z = ON)$ does not equal $P(y|I_Z = ON)$ for some Y-values y in every physically possible probability distribution P described by DEP'. If I_Z is a deterministic intervention variable, like in Woodward's (2003) original theory, then $I_Z = ON$ fixes Z's value and probabilistic influence of $I_X = ON$ on Y over the path $I_X \longrightarrow X \longrightarrow Z \longrightarrow Y$ is blocked. Thus, the theory would correctly tell us that X is not a direct cause of Y in S'. But if we accept Modification 5.3 and allow for stochastic intervention variables, then it may be that $I_Z = ON$ will not block the probabilistic influence of I_X on Y transported over the causal path $I_X \longrightarrow X \longrightarrow Z \longrightarrow Y$ and the two conditional probabilities $P(y|I_X = ON, I_Z = ON)$ and $P(y|I_Z = ON)$ may differ. In that case the theory would tell us that X is a direct cause of Y in S, what would contradict our initial assumption that X is only an indirect cause of Y.

The problem arises due to the fact that, when accepting Modification 5.3, it is no more guaranteed that interventions $I_Z = ON$ fix the values of the intervention variables' target variables Z. But, as demonstrated above, the fixability of all kinds of variables' values by means of interventions cannot simply be assumed. What one could do to avoid the problems is to modify Definition 5.8: One could simply give up the idea of fixability by interventions as a concept required for explicating the meaning of direct causation. Variables' values cannot only be fixed by interventions, but also by conditionalization. So what we have to do to avoid problematic cases such as the one illustrated in Fig. 5.3 without making use of interventions is to conditionalize on a suitably chosen set M of variables that blocks all indirect causal paths between X and Y. If Y depends on $I_X = ON$ conditional on such a blocking set M, then the I_X-Y dependence can only arise due to a direct causal connection $X \longrightarrow Y$. In directed acyclic graphs, M can be chosen as $Par(Y) \backslash \{X\}$.

According to the above considerations, one could improve the notion of direct causation further by also accepting the following modification in addition to Modification 5.3:

Modification 5.4 *Replace Definition 5.8 by the following definition:*

Direct causation: *If there exist weak i-expansions $S' = \langle V', E', DEP' \rangle$ of $S = \langle V, E, DEP \rangle$ for X w.r.t. Y, then $X \in V$ is a direct cause of Y in S if and only if $DEP'(Y, I_X = ON | Par(Y) \backslash \{X\})$, where I_X is an intervention variable for X w.r.t. Y in S'.*

Modification 5.4 suggests, like Modification 5.2, to characterize direct causation by a partial definition. In addition, it suggests to drop the fixability condition, which allows appliance of the interventionist theory to scenarios featuring variables one cannot fully control by interventions. (Thus, a weak i-expansion featuring a suitable intervention variable for X w.r.t. Y suffices for direct causation.)

With Modification 5.4 we now have a tool to pick out a variable Y's causal parents as direct causes in systems in which only stochastic interventions are available. But the notion of contributing causation may still make trouble in some systems. Assume we are interested in a system $S = \langle V, E, DEP \rangle$ with $V = \{X, Y, Z\}$ and causal structure $X \longrightarrow Y \longrightarrow Z$, where X and Y are dependent. Then we would like our definition of contributing causation to tell us that X is a contributing cause of Y. But now assume that Z's value cannot be fixed by an intervention variable in any strong i-expansion of S. Then our partial definition would not even be applicable to S. This observation motivates a modification of the notion of contributing causation that, similar to Modification 5.4 for direct causation, replaces the fixability condition by a conditionalization condition:

Modification 5.5 *Replace Definition 5.9 by the following definition:*

Contributing causation: *If every path between $X \in V$ and $Y \in V$ in $S = \langle V, E, DEP \rangle$ can be isolated and there exist weak i-expansions $S' = \langle V', E', DEP' \rangle$ of S for X w.r.t. Y, then $X \in V$ is a contributing cause of Y in S if and only if*

(a) *there is a chain π of direct causal relations (as defined above) from X to Y in S, and*

(b) *$DEP'(Y, I_X = ON | M)$ holds in some weak i-expansions, where M blocks all paths between X and Y different from π.*

Note that using stochastic intervention variables to account for contributing causation requires that all paths between a system's target and test variable of interest can be isolated. (Recall our findings in Sect. 5.2.2.) Because of this, the if-condition of the partial definition suggested in Modification 5.5 has to be stronger than the if-conditions of the stochastic versions of total and direct causation suggested above.

Let me briefly summarize the findings of this subsection. First, we observed that there are many variables which cannot be manipulated by physically possible interventions. In addition, there are also variables that even lack conceivable logically (or conceptually) possible interventions. To avoid false consequences in such cases, one has to modify Woodward's (2003) interventionist theory. The most obvious way to make the theory applicable to the problematic scenarios is to drop the only-if part of Definitions 5.7, 5.8, and 5.9. However, this move would deviate drastically from Woodward's intention to explicate causation. So one rather should accept Modification 5.2 and replace Woodward's original explicit definitions by partial definitions whose if-conditions require the existence of suitable intervention expansions. But then there are still scenarios containing variables whose values cannot be fully determined by interventions, which is required by condition (b) in Definition 5.6. To make the theory applicable also to these scenarios, one has to weaken condition (b) in Definition 5.6: One has to replace Woodward's original deterministic interventions by so-called stochastic interventions (Modification 5.3). This move is fine for total causation, but leads to new problems with the concept of direct causation, because accepting Modification 5.3 means dropping the fixability condition implemented in Definition 5.8: For correct inference to direct causal connections between variables X and Y it is required to block all indirect causal paths between X and Y, what cannot be done by soft interventions. However, these paths can be blocked by conditionalization (Modification 5.4). We can account for contributing causation in systems featuring variables which can only be manipulated by means of stochastic interventions by a similar conditionalization procedure (Modification 5.5). So Modification 5.2 should be accepted only for the notion of total causation, Modification 5.3 should be accepted for total, direct, and contributing causation, Modification 5.4 may be accepted for direct causation, and Modification 5.5 may be accepted for contributing causation.

5.3.2   Problem (ii): Interventions as Intermediate Causes

The second problem I want to discuss arises for Woodward's (2003) theory of causation in certain kinds of causal chains. Responsible for this problem is the fact that Woodward's definition of an intervention variable does not what it is intended to do, *viz.* picking out the causes as intervention variables that allow for correct causal inference according to Definitions 5.7, 5.8, and 5.9. This means that sometimes Definition 5.6 may pick out causes as intervention variables that lead, according to Definitions 5.7, 5.8, and 5.9, to false consequences.

Let me now illustrate the mentioned problem.[11] (See Fig. 5.4 for a graphical illustration.) Suppose we are interested in a causal system $S = \langle V, E, DEP \rangle$ with variable set $V = \{X, Y\}$ and causal structure $Y \longrightarrow X$. In this causal system X is not a cause of Y. However, it may be the case that there exists a strong i-expansion

---

[11]In illustrating the problem I use the notion of direct causation as a proxy for total, direct, and contributing causation. But nothing hinges on the use of the concept of direct causation here; the problem can also be formulated in terms of total or contributing causation.

(a)   (b)

Fig. 5.4 (a) shows our model $S = \langle V, E, DEP \rangle$ and (b) S's strong i-expansion $S' = \langle V', E', DEP' \rangle$ introducing an intervention variable I_X for X w.r.t. Y as an intermediate cause. It may now turn out that Y depends on $I_X = ON$. In that case, Definition 5.8 would imply the false consequence that X is a direct cause of Y

$S' = \langle V', E', DEP' \rangle$ of S w.r.t. Y with causal structure $Y \longrightarrow I_X \longrightarrow X$ in which I_X turns out to be an intervention variable for X w.r.t. Y. Let us see under which conditions this can be the case.

For I_X to be an intervention variable for X w.r.t. Y the four intervention conditions (a)–(d) of Definition 5.6 have to be satisfied. Condition (a) requires I_X to be a cause of X in S', which actually is the case in our i-expansion S'. Condition (b) is satisfied when I_X has some values i_X such that X depends on these values i_X and is forced to take a certain value x when I_X takes one of these values i_X. This may be true—why not? The third intervention condition requires that all directed causal paths from I_X to Y in S' go through X. Since there are no such directed causal paths from I_X to Y in S', this is trivially true and (c) is satisfied. To also satisfy the last condition, condition (d), I_X has to be independent of all those Z in any expansion of S' that cause Y over a causal path that does not go through X in the respective expansion. This requires that there are no common causes Z of I_X and Y in any expansion of S' or that every such common cause path is unproductive, and this, of course, can also be the case.

For further illustration, assume X stands for whether the burglar alarm is activated, while Y stands for the number of burglaries in the neighborhood in the last six months. Clearly the number of recent burglaries in the neighborhood is causally relevant for whether the burglar alarm is activated, and not the other way round. Now assume I_X to represent one's decision to activate the burglar alarm. Condition (a) is satisfied because one's decision to activate the burglar alarm is clearly a cause for whether the burglar alarm is activated. In addition, the number of recent burglaries can also be expected to have an effect on one's decision to activate the burglar alarm. If we prescind from malfunctions etc., whether the burglar alarm is activated will deterministically depend on one's decision to activate the burglar alarm. Hence, also condition (b) will be satisfied. It seems also plausible that conditions (c) and (d) are satisfied: The number of burglaries in the neighborhood in the last six months can cause the activation of the burglar alarm only by causing one to decide to activate the alarm, and there are—at least at first glance—no plausible candidates for common causes of number of recent burglaries in the neighborhood and activation of the burglar alarm.

What we have done so far is that we have constructed a hypothetically possible causal scenario with causal structure $Y \longrightarrow I_X \longrightarrow X$ in which I_X is an intervention variable for X w.r.t. Y. Now it may also be the case that I_X and Y are dependent in such a scenario (what is quite plausible in general in causal chains and can be expected to be the case in our burglar alarm example). Thus, Definition 5.8 would imply that X has to be a direct cause of Y in S, what contradicts our initial assumption that X is not a cause of Y in S. Applied to our example Definition 5.8 leads to the consequence that whether the burglar alarm is activated is causally relevant for the number of burglaries in the neighborhood in the last six months, which is obviously false.

Note that the problem described above does not only arise for Woodward's (2003) original interventionist theory of causation, but also for Woodward's theory plus any possible combination of modifications discussed in Sect. 5.3.1. In all versions of Woodward's theory it is not excluded that there are i-expansions featuring intervention variables I_X for X w.r.t. Y that are correlated with Y as intermediate causes. This would always lead to the false consequence that X is a direct cause of Y.

The most promising way to avoid the problem described above seems to consist in adding a condition to the definition of an intervention variable (Definition 5.6) that narrows the extension of the notion of an intervention variable in such a way that exactly the problematic variables, such as I_X in our exemplary causal chain $Y \longrightarrow I_X \longrightarrow X$, are not allowed to be intervention variables for X w.r.t. Y anymore. So for I_X to be an intervention variable for X w.r.t. Y it should be required that I_X is exogenous:

Modification 5.6 *Add the following as a necessary condition for I_X to be an intervention variable for X w.r.t. Y in $\langle V, E, DEP \rangle$ to Definition 5.6: I_X is exogenous in S.*

5.3.3   Problem (iii): Deterministic Causal Chains and Deterministic Common Cause Structures

The third problem I want to discuss in this section arises for certain deterministic causal systems. In particular, it arises for direct and contributing causation in causal systems containing physically deterministic causal chains of minimum length three and for physically deterministic common cause structures. By a physically deterministic system I mean a causal system whose set of variables contains at least one physically (or nomologically) deterministic cause. A *physically deterministic cause* is a cause that fully controls the value of its effect in every physically possible world, meaning that in every physically possible world for every possible value x of a physically deterministic cause X there is exactly one value y of its effect Y such that Y is forced to take this value y when $X = x$. Here comes the definition of physically deterministic causation I will use in the remainder of this chapter[12]:

---

[12]I will most of the time simply speak of a deterministic cause when meaning a physically (or nomologically) deterministic cause.

Definition 5.10 (deterministic cause) X is a (physically) deterministic cause of Y in $S = \langle V, E, DEP \rangle$ if and only if

(a) $X \longrightarrow\longrightarrow Y$ in S, and
(b) for every $\langle V, E \rangle$-physically possible probability distribution P and every X-value x there is an Y-value y such that $P(y|x) = 1$ holds in every context $C = c$.

Condition (a) says that X is a cause of Y. Condition (b) guarantees that X nomologically determines Y; X's deterministic infleunce on Y cannot be broken by changing the context from $C = c$ to some value $c' \neq c$ in any physically possible distribution P.

Definition 5.10 implies, given our interventionist framework, that it is impossible to directly manipulate the effects of a deterministic cause:

Theorem 5.1 *If $S = \langle V, E, DEP \rangle$ is a causal dependence model, $X, Y \in V$, and X is a deterministic cause of Y in S, then there is no intervention variable I_Y for Y in any strong i-expansion $S' = \langle V', E', DEP' \rangle$ of S for any $Z \in V$ with $Z \neq Y$.*

Proof Assume $S = \langle V, E, DEP \rangle$ is a causal dependence model, $X, Y, Z \in V$, and X is a deterministic cause of Y in S. Assume further that there is a strong i-expansion $S' = \langle V', E', DEP' \rangle$ of S w.r.t. Z such that $I_Y \in V'$ is an intervention variable for Y w.r.t. Z in S'. Then I_Y is a cause of Y in S'. For I_Y to turn out as a direct or contributing cause of Y there has to be a strong i-expansion $S'' = \langle V'', E'', DEP'' \rangle$ of S' w.r.t. Y containing an intervention variable I_{I_Y} for I_Y w.r.t. Y and an intervention variable I_X for X w.r.t. Y such that $DEP''(Y, I_{I_Y} = ON|I_X = ON)$ holds. So there has to be a $\langle V'', E'' \rangle$-physically possible probability distribution P (represented by DEP'') such that $P(y, I_{I_Y} = ON|I_X = ON) \neq P(y|I_X = ON)$ holds for some Y-values y. According to Definition 5.6(b), there is exactly one X-value x such that $P(x|I_X = ON) = 1$. Since X is a deterministic cause of Y, it follows from Definition 5.10(b) that there is exactly one Y-value y with $P(y|x) = 1$. Since conditional probabilities of 1 are transitive, it follows that $P(y|I_X = ON) = 1$ and that $P(y'|I_X = ON) = 0$ for all other Y-values y'. Since conditional probabilities of 1 and 0 cannot be changed by conditionalizing on additional variables' values, $P(y|I_{I_Y} = ON, I_X = ON) = P(y|I_X = ON)$ holds for all Y-values y, what contradicts the earlier consequence that $P(y|I_{I_Y} = ON, I_X = ON) \neq P(y|I_X = ON)$ holds for some Y-values y. □

Why and how does this theorem lead to problems for Woodward's (2003) original interventionist theory of causation? As mentioned above, a problem does arise in systems whose causal structures feature deterministic causal chains of minimum length three or deterministic common cause paths. A deterministic *causal chain* is a causal path of the form $X_1 \longrightarrow \ldots \longrightarrow X_n$ such that every X_i (with $1 \leq i < n$) is a deterministic cause of every X_j with $i < j \leq n$. A deterministic *common cause path* is a common cause path $X \longleftarrow\longleftarrow Z \longrightarrow\longrightarrow Y$ such that $Z \longrightarrow\longrightarrow X$ as well as $Z \longrightarrow\longrightarrow Y$ are deterministic causal chains. Let me now demonstrate the problem on an example of a deterministic causal chain of minimum length three and on an example of a deterministic common cause structure.

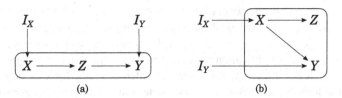

Fig. 5.5 Since Y's value cannot be fixed by means of an intervention variable I_Y for Y in any expansion of a causal system featuring (**a**) a deterministic causal chain of minimum length three or (**b**) a deterministic common cause structure, Definition 5.8's definiens cannot be satisfied when X is a direct cause of Z in S. Thus, the interventionist theory of causation leads to the false result that X is not a direct cause of Z in both cases

I begin with the deterministic causal chain. Assume we are interested in a system $S = \langle V, E, DEP \rangle$ with $V = \{X, Y, Z\}$ and causal structure $X \longrightarrow Z \longrightarrow Y$, where $X \longrightarrow Z \longrightarrow Y$ is a deterministic causal chain. (See Fig. 5.5a for a graphical illustration.) We now ask Woodward's (2003) interventionist theory of causation whether X is a direct cause of Z in this causal scenario.[13] The theory tells us (Definition 5.8) that X is a direct cause of Z in S if and only if there is a strong i-expansion $S' = \langle V', E', DEP' \rangle$ of S w.r.t. Z whose variable set V' contains an intervention variable I_X for X w.r.t. Z as well as an intervention variable I_Y for Y w.r.t. Z such that Z depends on $I_X = ON$ when $I_Y = ON$. But because Y is an effect of a deterministic cause, *viz.* of X, Theorem 5.1 implies that there cannot be such an intervention variable I_Y for Y in any strong i-expansion of S w.r.t. Z. Thus, Definition 5.8's definiens cannot be satisfied w.r.t. X and Z, and hence, X cannot be a direct cause of Z in S.

The same problem arises for deterministic common cause structures. Assume $S = \langle V, E, DEP \rangle$ to be a causal system with variable set $V = \{X, Y, Z\}$ and causal structure $Z \longleftarrow X \longrightarrow Y$, where X is a deterministic cause of Y and of Z. (See Fig. 5.5b for a graphical illustration.) Again we ask our interventionist theory of causation about the causal relationship between X and Z: Is X a direct cause of Z in S? And, again, the theory (in particular Definition 5.8) tells us that X is a direct cause of Z in S if and only if the system can be expanded in such a way that the expansion features an intervention variable I_X for X w.r.t. Z and an intervention variable I_Y for Y w.r.t. Z such that $I_X = ON$ has some influence on Z when $I_Y = ON$. But, like in the deterministic chain case, Y is an effect of a deterministic cause, *viz.* of X. With Theorem 5.1 it follows then that there cannot be such an intervention variable I_Y for Y w.r.t. Z in any i-expansion, and thus, Definition 5.8 tells us that X is not a direct cause of Z in S.

First of all, note that the problem does not arise for total causation, but only for direct and contributing causation. It arises due to the fixability condition implemented in Definitions 5.8 and 5.9. Now there are several ways of how the problem described in this subsection can be solved. I will discuss four of these

---

[13] The same reasoning can be done for the notion of contributing causation.

possibilities. The first one would be to accept Modification 5.1: Drop the only-if part of Definitions 5.8 and 5.9. We already saw in Sect. 5.3.1 that accepting Modification 5.1 does not really provide an explication of causation anymore. Another inconvenience with Modification 5.1 is that it would, though avoiding the problem illustrated in this subsection, not allow for inference of effects of deterministic causes in deterministic causal chains and deterministic common cause structures. The reason for this is simply (as demonstrated above) that Definitions 5.8's and 5.9's definiens cannot be satisfied in such causal scenarios.

What is with accepting Modification 5.2? It suggests to replace Definitions 5.8 and 5.9 by partial definitions whose if-conditions require the existence of suitable intervention variables. Thus, when accepting Modification 5.2, our modified interventionist notions of direct and contributing causation are not applicable to deterministic causal chains and deterministic common cause structures anymore—they keep silent about direct and contributing causation in such systems.

If we do want to explicate direct and contributing causation in deterministic systems as described in this subsection by means of interventions, we should replace Definitions 5.8 and 5.9 by partial definitions based on stochastic instead of deterministic interventions, i.e., we should decide in favor of Modifications 5.4 and 5.5. The stochastic version of Woodward's interventionist theory does not require fixability of all off-path variables $Y \in V$ different from the intervention variable I_X's target variable X and test variable Z. So the problem described in this subsection does not arise for the stochastic version.

Another way to go seems to consist in following Woodward's proposal of a modified interventionist theory in Woodward (2015). In this paper Woodward is interested in the question of how his interventionist theory of causation can be successfully applied to causal systems whose sets of variables contain variables which stand in relationships of supervenience, or, more generally, in relationships of non-causal dependence. Let me first demonstrate why Woodward's (2003) original theory leads to problems quite similar to the ones we get with deterministic systems when applied to such systems involving non-causal dependencies. Assume we are interested in a causal system $S = \langle V, E, DEP \rangle$ with variable set $V = \{I_M, M, P\}$. M stands for a mental property that supervenes on a physical property P. I_M should be an intervention variable for M. Note that I_M has to be a causal parent of M in S in that case. If we want to account for the causal relation between I_M and M in terms of the interventionist theory, there has to be a strong i-expansion $S' = \langle V', E', DEP' \rangle$ of S w.r.t. M such that V' contains an intervention variable I_{I_M} for I_M w.r.t. M as well as an intervention variable I_P for P w.r.t. M such that M depends on $I_{I_M} = ON$ given $I_P = ON$. But, since M supervenes on P, every change in M is necessarily accompanied by a change in P (cf. McLaughlin and Bennett 2011, sec. 1), i.e., M cannot be changed without changing P. But this means that M is independent of $I_{I_M} = ON$ conditional on $I_P = ON$, and thus, that I_M cannot be a direct cause of M in S. Hence, I_M cannot be an intervention variable for M in S. (See also Fig. 5.6.)

As already mentioned above, the problem arises due to the fixability condition implemented in Definition 5.8 (and likewise for Definition 5.9): For I_M to be a direct cause of M there has to be an intervention variable I_{I_M} that has some influence

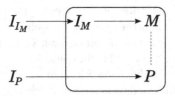

Fig. 5.6 The *dashed line* stands for a supervenience relationship between M and P. According to Woodward's (2003) interventionist theory, there cannot be an intervention variable I_M for M in any system whose variable set contains both M and P. Such an intervention variable I_M would have to be a causal parent of M. But then it would follow from Definition 5.8 that there is also an intervention variable I_{I_M} for I_M w.r.t. M that would change M's probability distribution when P's value is fixed by another intervention variable I_P. However, since M supervenes on P, P's value cannot be fixed by any such intervention variable I_P for P when M's value is changed by an intervention variable I_{I_M} for I_M w.r.t. M

on M when P's value is fixed, i.e., in contexts $I_P = ON$. Woodward (2015, sec. 6) decides to loosen the fixability condition implemented in his definition of direct and contributing causation to allow for intervention variables I_M for M in systems featuring supervenience relationships. In particular, he drops the fixability requirement for those variables of the system of interest's variable set that stand in supervenience relationships to the variables whose causal connection should be tested.

Our problem with deterministic causal chains of minimum length three and deterministic common cause structures is quite similar to Woodward's (2015) problem: Our problem arises because in deterministic structures, such as $X \longrightarrow Z \longrightarrow Y$ and $Z \longleftarrow X \longrightarrow Y$ depicted in Fig. 5.5, there cannot be causes of (and hence no intervention variables for) Y that cause Y over a path not going through X. But the existence of such intervention variables, by whose means Y's value could be fixed, is required for X to be a direct cause of Z. In analogy to Woodward's (2015) proposal to drop the fixability condition for variables standing in supervenience relationships, we could accept a modification that suggests to drop the fixability condition for variables standing in relationships of deterministic causal dependence. This would amount to accepting the following modification:

Modification 5.7 *Replace 'I$_Z$ is the set of intervention variables for variables $Z \in V$ (different from X and Y) w.r.t. Y in S'' in Definition 5.8 by 'I$_Z$ is the set of intervention variables for variables $Z \in V$ w.r.t. Y in S' which do not deterministically depend on X or Y'.*

Replace 'I$_Z$ is the set of all intervention variables for variables $Z \in V$ (not lying on path π) w.r.t. Y in S'' in Definition 5.9 by 'I$_Z$ is the set of intervention variables for variables $Z \in V$ (not lying on path π) w.r.t. Y in S' which do not deterministically depend on X or Y'.

Let us now see how accepting this latest modification can avoid the problem by means of the two deterministic exemplary causal structures $X \longrightarrow Z \longrightarrow Y$ and $Z \longleftarrow X \longrightarrow Y$. The problem arose when we asked Woodward's (2003) original theory whether X is a direct cause of Z in these systems. Because, according to Definition 5.8, for X to be a direct cause of Z requires an intervention variable I_Y and Theorem 5.1 tells us that such an intervention variable cannot exist for effects of deterministic causes, the theory led to the consequence that X is not a direct cause of Z. But when we accept Modification 5.7, such an intervention variable I_Y is not anymore required for X to be a direct cause of Z when Y stands in a deterministic relationship to X. Thus, Definition 5.8 together with Modification 5.7 leads to the correct result that X is a direct cause of Z.

But though the initial problem with deterministic causal chains of minimum length three and deterministic common cause structures can be avoided when accepting Modification 5.7, there is a new problem we have to face. Let me illustrate this new problem, again, by means of the exemplary deterministic causal chain $X \longrightarrow Z \longrightarrow Y$. In this causal structure X is not a causal parent of Y. But when accepting Modification 5.7, Definition 5.8 tells us that X is a direct cause of Y if and only if the system can be expanded in such a way that this expansion contains an intervention variable I_X for X w.r.t. Y such that Y's probability distribution is changed when $I_X = ON$ and all variables not deterministically dependent on X or Y are fixed by intervention variables. In our example there are no such variables, and thus, no additional intervention variables are required for checking whether X is a direct cause of Y. What is required for X to be a direct cause of Y is, hence, only the existence of an intervention variable I_X for X w.r.t. Y in a strong i-expansion $S' = \langle V', E', DEP' \rangle$ of the system of interest such that $DEP'(Y, I_X)$. There may, of course, be such an intervention variable. Hence, it may turn out that X is a direct cause of Y, what would contradict our initial assumption that X is not a causal parent of Y.

Summarizing, Woodward's (2003) interventionist theory of causation leads to problems in deterministic causal chains of minimum length three and in deterministic common cause structures. The problem with such systems is that direct causal dependence requires that all variables beside the two which are tested for direct causal dependence have to be fixed, what is impossible in systems with deterministic causes that have more than one effect. (A similar problem arises for contributing causal dependence.) I have discussed four possible modifications one could think of to avoid the problem: One could (i) accept Modification 5.1. The disadvantages of this move are that it does not allow for inferring a deterministic cause's effects in the discussed problematic scenarios and that it does not help much when one wants, as Woodward does, to explicate causation in terms of interventions. So is it a better move to (ii) accept Modification 5.2? This move avoids the problem discussed, but has the disadvantage that the modified interventionist theory would be silent in case of the problematic systems simply because the if-conditions of the partial definitions of direct and contributing causation would not be satisfied in such systems. The best way to go seems to be to (iii) accept Modifications 5.4 and 5.5. This allows for inferring effects of deterministic common causes as well as for stochastic interventions in the sense of Modification 5.3. A fourth possibility seems to be to (iv) accept something like Modification 5.7. This solves the initial problem discussed in this subsection, but, as we saw, directly leads to new problems.

5.4 An Alternative Approach

In this section I want to do two things. First (Sect. 5.4.1), I will present an alternative version of Woodward's (2003) interventionist theory of causation that is, like Woodward's original theory, based on the notion of deterministic interventions. This alternative version will be the result of accepting the most reasonable combination of modifications for deterministic interventions discussed in Sect. 5.3. After introducing this alternative theory, I will demonstrate that it does not have to be seen as an independent theory of causation, but neatly follows from assuming that certain combinations of the causal Markov condition (CMC), the causal minimality condition (Min), and the causal faithfulness condition (CFC) hold for suitable intervention expansions. Second (Sect. 5.4.2), I will extend the main results of Sect. 5.4.1 to systems for which only stochastic interventions are available. In the whole section we assume all causal structures considered to be acyclic.

5.4.1 An Alternative Theory Using Deterministic Interventions

Given that we do not want to use stochastic interventions, there is a deficiency w.r.t. Woodward's (2003) theory due to his original definition of an intervention variable. This definition lacks of a condition requiring intervention variables to be exogenous (see Sect. 5.3.2). When adding this missing requirement as a condition (Modification 5.3) to condition (a) in Definition 5.6, we arrive at the following improved definition of a deterministic intervention variable:

Definition 5.11 (intervention variable^D) If $S = \langle V, E, DEP \rangle$ is a causal dependence model with $I_X, X, Y \in V$, then I_X is an intervention variable for X w.r.t. Y in S if and only if

(a) $I_X \longrightarrow X$ in S and I_X is exogenous in S, and
(b) for every $\langle V, E \rangle$-physically possible probability distribution P over V represented by DEP and every $i_X \in ON$ there is an $x \in val(X)$ such that $DEP_P(x, i_X)$ and $P(x|i_X) = 1$, and
(c) all paths $I_X \longrightarrow\longrightarrow Y$ in S go through X, and
(d) $INDEP(I_X, Z)$ holds for all Z that cause Y over a directed path that does not go through X in any expansion $S' = \langle V', E', DEP' \rangle$ of S.

When insisting on using deterministic interventions, then the only problem with Woodward's (2003) notions of total, direct, and contributing causation we discussed in Sect. 5.3 was that the original definitions (Definitions 5.7, 5.8, and 5.9, respectively) do not always pick out the causal paths $X \longrightarrow\longrightarrow Y$ they should. In particular, Definition 5.7 applied to a system $S = \langle V, E, DEP \rangle$ with $X, Y \in V$ and $X \longrightarrow\longrightarrow Y$ in S satisfying CMC and CFC such that there is no physically possible deterministic intervention on X w.r.t. Y, would lead to the result that X is not a total cause of Y. If S satisfies CMC and Min, $X \longrightarrow Y$ in S, and X cannot be influenced by deterministic interventions, the original definition of direct causation falsely tells us that X is not a direct cause of Y. Finally, if S would satisfy CMC and every directed causal path $\pi : X \longrightarrow\longrightarrow Y$ in S could be isolated and tested for productiveness, then X should turn out to be a contributing cause of Y in S if and only if there is at least one productive directed path $\pi : X \longrightarrow\longrightarrow Y$ in S. But if X cannot be manipulated by deterministic interventions, then the original definition of contributing causation tells us that X is not a contributing cause of Y even if there is such a productive path π.

The best way to solve the problem for all three notions of causation discussed in Sect. 5.3 turned out to be accepting Modification 5.2, i.e., replacing the original explicit definitions of total, direct, and contributing causation by partial definitions whose if-conditions require the existence of suitable interventions. Following Modification 5.2 we arrive at the following improved definitions of causation*:

Definition 5.12 (total causation^D) If there exist weak i-expansions $S' = \langle V', E', DEP' \rangle$ of $S = \langle V, E, DEP \rangle$ for X w.r.t. Y, then X is a total cause of Y in S if and only if $DEP'(Y, I_X = ON)$ holds in some weak i-expansions, where I_X is an intervention variable for X w.r.t. Y in S'.

Definition 5.13 (direct causation^D) If there exist strong i-expansions $S' = \langle V', E', DEP' \rangle$ of $S = \langle V, E, DEP \rangle$ w.r.t. Y, then $X \in V$ is a direct cause of Y in S if and only if $DEP'(Y, I_X = ON | I_Z = ON)$ holds in some strong i-expansions, where I_X is an intervention variable for X w.r.t. Y in S' and I_Z is the set of all intervention variables in S' different from I_X.

Definition 5.14 (contributing causation^D) If there exists a strong i-expansions $S' = \langle V', E', DEP' \rangle$ of $S = \langle V, E, DEP \rangle$ w.r.t. Y, then $X \in V$ is a contributing cause of Y in S if and only if

(a) there is a chain π of direct causal relations (in the sense of Definition 5.13) from X to Y in S, and

(b) $DEP'(Y, I_X = ON | I_Z = ON)$ holds in some strong i-expansions, where I_X is an intervention variable for X w.r.t. Y in S' and I_Z is the set of intervention variables for variables in V not lying on π.

Next I show that these three improved definitions actually pick out the causal chains $X \longrightarrow\longrightarrow Y$ they should in systems satisfying the respective pre-requirements mentioned above. I start with total causation. In systems satisfying CMC and CFC for which suitable i-expansions exist Definition 5.12 determines X to be a total cause of Y if and only if X is a cause of Y:

Theorem 5.2 *If for all $X, Y \in V$ (with $X \neq Y$) there exist weak i-expansions $S' = \langle V', E', DEP' \rangle$ of $S = \langle V, E, DEP \rangle$ for X w.r.t. Y satisfying CMC and CFC, then for all $X, Y \in V$ (with $X \neq Y$): $X \longrightarrow\longrightarrow Y$ in S if and only if $DEP'(Y, I_X = ON)$ holds in some weak i-expansions S', where I_X is an intervention variable for X w.r.t. Y in S'.*

Proof Suppose $S = \langle V, E, DEP \rangle$ to be a causal dependence model. Suppose further that for every $X, Y \in V$ (with $X \neq Y$) there exist weak *i*-expansions of S for X w.r.t. Y satisfying CMC and CFC. Now let X and Y be arbitrarily chosen elements of V with $X \neq Y$. Let I_X be an intervention variable for X w.r.t. Y in S', where S' is one of the presupposed weak *i*-expansions for X w.r.t. Y.

The if-direction: Assume that $DEP'(Y, I_X = ON)$ holds in S'. Then the *d*-connection condition implies that there is a causal path π *d*-connecting I_X and Y. π cannot have the form $I_X \longleftarrow \ldots - Y$. This is excluded by condition (a) of Definition 5.11. So π must have the form $I_X \longrightarrow \ldots - Y$. π cannot feature colliders. Otherwise π would be blocked by the empty set and π would not *d*-connect I_X and Y. Hence, π must be a directed path $I_X \longrightarrow\longrightarrow Y$. Now either (A) π goes through X, or (B) π does not go through X. Since (B) is excluded by Definition 5.11(c), (A) must be the case. Then, since S' is a weak *i*-expansion of S, there is also a directed causal path $X \longrightarrow\longrightarrow Y$ in S.

The only-if-direction: Suppose $X \longrightarrow\longrightarrow Y$ in S. Then $X \longrightarrow\longrightarrow Y$ is also in S'. Since I_X is an intervention variable for X, there is also a causal path $I_X \longrightarrow X$ in S'. The concatenation of this path $I_X \longrightarrow X$ and $X \longrightarrow\longrightarrow Y$ is $I_X \longrightarrow X \longrightarrow\longrightarrow Y$. With faithfulness now $DEP'(Y, I_X)$ follows. Since $I_X = OFF$ is not correlated with any variable in V, Y must be dependent on $I_X = ON$, i.e., $DEP(Y, I_X = ON)$.   □

Next I show that Definition 5.13 determines X to be a direct cause of Y if and only if X is a causal parent of Y. This result does only hold in systems satisfying CMC and Min for which suitable *i*-expansions exist. Note the connection to Zhang and Spirtes (2011), who show vice versa that CMC together with an interventionist definition of direct causation implies Min.

The following proof is based on a proof in (Gebharter and Schurz 2014, sec. 5) adapted for dependence models.

Theorem 5.3 *If for every $Y \in V$ there exist strong i-expansions $S' = \langle V', E', DEP' \rangle$ of $S = \langle V, E, DEP \rangle$ w.r.t. Y satisfying CMC and Min, then for all $X, Y \in V$ (with $X \neq Y$): $X \longrightarrow Y$ in S if and only if $DEP'(Y, I_X = ON | I_Z = ON)$ holds in some strong i-expansions S', where I_X is an intervention variable for X w.r.t. Y in S' and I_Z is the set of all intervention variables in S' different from I_X.*

Proof Suppose $S = \langle V, E, DEP \rangle$ is a causal dependence model. Suppose further that for every $Y \in V$ there are strong *i*-expansions of S w.r.t. Y satisfying CMC and Min. Now let X and Y be arbitrarily chosen elements of V with $X \neq Y$. Let I_X be an intervention variable for X w.r.t. Y in S', and let I_Z be the set of all intervention variables in S' different from I_X (where S' is one of the presupposed strong *i*-expansions of S w.r.t. Y).

The if-direction: Assume that $DEP'(Y, I_X = ON | I_Z = ON)$ holds. Then $DEP'(Y, I_X = ON | I_Z = ON)$ together with the *d*-connection condition implies that I_X and Y are *d*-connected over a causal path π. This path π cannot be a collider path, because then I_X and Y would be *d*-separated over π. But due to Definition 5.11(a) π can also not have the form $I_X \longleftarrow \ldots - Y$. Thus, π must have the form $I_X \longrightarrow \ldots \longrightarrow Y$. Since π cannot be a collider path, π must be a directed path $I_X \longrightarrow\!\!\!\longrightarrow Y$. Now there are two possible cases: Either (A) π is going through X, or (B) π is not going through X. (B) is excluded by Definition 5.11(c). Hence, (A) must be the case. If (A) is the case, then either (i) $X \longrightarrow Y$ is part of π, or (ii) $X \longrightarrow\!\!\!\longrightarrow C \longrightarrow\!\!\!\longrightarrow Y$ is part of π.

Assume (ii) is the case. Let r be an individual variable ranging over $val(Par(Y))$. Let P be a probability distribution over V' whose (in)dependencies are represented by DEP'. Let $P^*(-)$ be defined as $P(-|I_Z = ON)$. We proceed by stating the following probabilistically valid equations:

$$P^*(y|I_X = ON) = \sum_r P^*(y|r, I_X = ON) \cdot P^*(r|I_X = ON) \tag{5.1}$$

$$P^*(y) = \sum_r P^*(y|r) \cdot P^*(r) \tag{5.2}$$

$I_Z = ON$ forces all non-intervention variables in V' to take certain values. Hence, $I_Z = ON$ also forces $Par(Y)$ to take a certain value r^*. This means that $P^*(r^*) = 1$ and that $P^*(r) = 0$ for every $r \neq r^*$. Since probabilities of 1 cannot be changed by conditionalization, we also get $P^*(r^*|I_X = ON) = 1$ and $P^*(r|I_X = ON) = 0$ for every $r \neq r^*$. Now we get from Equations 5.1 and 5.2:

$$P^*(y|I_X = ON) = P^*(y|r^*, I_X = ON) \cdot 1 \tag{5.3}$$

$$P^*(y) = P^*(y|r^*) \cdot 1 \tag{5.4}$$

Since $Par(Y)$ blocks π, we get $P^*(y|r^*, I_X = ON) = P^*(y|r^*)$ with the d-connection condition. From this and Equations 5.3 and 5.4 it follows that $P^*(y|I_X = ON) = P^*(y)$. This means that $I_X = ON$ and Y are independent in P conditional on $I_Z = ON$ (i.e., $INDEP'(Y, I_X = ON|I_Z = ON)$), what contradicts the initial assumption that $DEP'(Y, I_X = ON|I_Z = ON)$ holds. Hence, (i) must be the case, i.e., π must have the form $I_X \longrightarrow\!\!\!\longrightarrow X \longrightarrow Y$. Since S' is a strong i-expansion of S, $X \longrightarrow Y$ in S.

The only-if-direction: Suppose $X \longrightarrow Y$ in S. From S' being an i-expansion of S it follows that $X \longrightarrow Y$ in S'. Because Min is equivalent to Prod, it follows that $DEP'(y, x|Par(Y)\backslash\{X\} = r)$ holds for some $x \in val(X)$, some $y \in val(Y)$, and some instantiations r of $Par(Y)\backslash\{X\}$. Since $Par(Y)\backslash\{X\}$ can also be forced to take value r by some ON-value of I_Z, also $DEP'(y, x|I_Z = ON)$ and $DEP'(y, x|I_Z = ON, r)$ hold. Let ON be one of the I_X-values that are correlated with x and that force X to take value x. (The existence of such a value ON is guaranteed by Definition 5.4.) We now have $DEP'(I_X = ON, x|I_Z = ON, r) \wedge DEP'(x, y|I_Z = ON, r)$. The axiom of weak union (see Sect. 2.5), which is probabilistically valid, gives us Equations 5.5 and 5.6 (in which $s = \langle x, r \rangle$ is a value realization of $Par(Y)$):

$$INDEP'(I_X = ON, s = \langle x, r \rangle | I_Z = ON) \Rightarrow INDEP'(I_X = ON, x | I_Z = ON, r) \tag{5.5}$$

$$INDEP'(s = \langle x, r \rangle, y | I_Z = ON) \Rightarrow INDEP'(x, y | I_Z = ON, r) \tag{5.6}$$

With the contrapositions of Equations 5.5 and 5.6 it now follows that $DEP'(I_X = ON, s = \langle x, r \rangle | I_Z = ON) \wedge DEP'(s = \langle x, r \rangle, y | I_Z = ON)$ holds.

Next, we show that $\exists s : DEP'(I_X = ON, s | I_Z = ON) \wedge DEP'(s, y | I_Z = ON)$ together with the d-connection condition implies $DEP'(I_X = ON, y | I_Z = ON)$. Let P be any probability distribution over V' whose (in)dependencies are represented by DEP'. Let $P^*(-)$ be defined as $P(-|I_Z = ON)$. We proceed by stating the following probabilistically valid equation:

$$P^*(y | I_X = ON) = \sum_i P^*(y | s_i, I_X = ON) \cdot P^*(s_i | I_X = ON) \tag{5.7}$$

Since $Par(Y)$ blocks all paths between I_X and Y, also the following equation holds:

$$P^*(y | I_X = ON) = \sum_i P^*(y | s_i) \cdot P^*(s_i | I_X = ON) \tag{5.8}$$

When $I_Z = ON$, $I_X = ON$ forces $Par(Y)$ to take value s. Hence, $P^*(s_i | I_X = ON) = 1$ in case $s_i = s$, and $P^*(s_i | I_X = ON) = 0$ otherwise. Thus, we get from Equation 5.8:

$$P^*(y | I_X = ON) = \sum_i P^*(y | s_i) \cdot 1 \tag{5.9}$$

For reductio, let us assume that $INDEP'(I_X = ON, y | I_Z = ON)$ holds, meaning that $P^*(y | I_X = ON) = P^*(y)$. Then we get the following equation from Equation 5.9:

$$P^*(y) = \sum_i P^*(y | s_i) \cdot 1 \tag{5.10}$$

Equation 5.10 implies $INDEP'(s, y | I_Z = ON)$, what contradicts $DEP'(s, y | I_Z = ON)$ above. Thus, $DEP'(I_X = ON, y | I_Z = ON)$ has to hold when $DEP'(I_X = ON, s | I_Z = ON) \wedge DEP'(s, y | I_Z = ON)$ holds. Therefore, $DEP'(I_X = ON, y | I_Z = ON)$. $\qquad\square$

The next theorem says that the definition of contributing causation (Definition 5.14) coincides with the existence of at least one directed causal path in systems with i-expansions that (i) satisfy CMC, (ii) in which every directed causal path can be isolated, and (iii) in which every two variables X and Y connected by a directed causal path are connected by at least one productive directed causal path:

Theorem 5.4 *If for all $Y \in V$ there exist strong i-expansions $S' = \langle V', E', DEP' \rangle$ of $S = \langle V, E, DEP \rangle$ w.r.t. Y that satisfy CMC, in which every directed causal path can be isolated, and in which for every directed path there is a productive directed path, then for all $X, Y \in V$ (with $X \neq Y$): There is a path $\pi : X \longrightarrow\!\!\!\!\longrightarrow Y$ in S if and only if*

(a) there is a chain π of direct causal relations (in the sense of Definition 5.13) from X to Y in S, and

(b) $DEP'(Y, I_X = ON | I_Z = ON)$ holds in some strong i-expansions S', where I_X is an intervention variable for X w.r.t. Y in S' and I_Z is the set of all intervention variables in S' for variables in V not lying on π.

Proof Assume that for every $Y \in V$ there exists a strong i-expansion $S' = \langle V', E', DEP' \rangle$ of $S = \langle V, E, DEP \rangle$ w.r.t. Y that satisfies CMC. In addition, suppose that for every directed path in S' there is a productive directed path in S' and that all directed causal paths in S' can be isolated. Let $X, Y \in V$ (with $X \neq Y$) be arbitrarily chosen.

The if-direction: The if-direction is trivial.

The only-if-direction: Assume There is a path $X \longrightarrow\!\!\!\!\longrightarrow Y$ in S. Since S' is a strong i-expansion of S, there is an intervention variable I_X for X w.r.t. Y in S'. Since I_X is an intervention variable for X w.r.t. Y, also the concatenation of $I_X \longrightarrow X$ and path $X \longrightarrow\!\!\!\!\longrightarrow Y$ is in S'. By assumption this implies the existence of a productive directed path $\pi : I_X \longrightarrow X \longrightarrow\!\!\!\!\longrightarrow Y$ in S'. Thus, $DEP'(Y, I_X | M)$ holds, where $M \subseteq V \backslash \{X, Y\}$ blocks all paths between X and Y different from π. Since M can be forced to take any value m by some ON-value of I_Z, there will also be some ON-value of I_Z such that $DEP'(Y, I_X = ON | I_Z = ON)$ holds. Since every arrow of path π is productive, there will also be a corresponding causal chain of directed causal relations (in the sense of Definition 5.13) from X to Y.   □

5.4.2 An Alternative Theory Using Stochastic Interventions

In this subsection I present an alternative version of Woodward's (2003) interventionist theory of causation based on stochastic interventions. This stochastic alternative variant of the interventionist theory is the result of accepting the most reasonable modifications discussed in Sect. 5.3 allowing for stochastic intervention variables. For Woodward's notion of an intervention variable this means to accept Modification 5.3 as well as Modification 5.6. Modification 5.3, which suggests to drop the requirement that interventions have to be deterministic, is implemented in condition (b) of Definition 5.15, and Modification 5.6, which requires intervention variables to be exogenous, is realized in condition (a) of Definition 5.15.[14]

---

[14]Note that accepting these modifications means that every intervention variable^D is an intervention variable^S.

Definition 5.15 (intervention variable^S) If $S = \langle V, E, DEP \rangle$ is a causal dependence model with $I_X, X, Y \in V$, then I_X is a stochastic intervention variable for X w.r.t. Y in S if and only if

(a) $I_X \longrightarrow X$ in S and I_X is exogenous in S, and
(b) for every $\langle V, E \rangle$-physically possible probability distribution P over V represented by DEP and every $i_X \in ON$ there is an $x \in val(X)$ such that $DEP_P(x, i_X)$, and
(c) all paths $I_X \longrightarrow\longrightarrow Y$ in S go through X, and
(d) $INDEP'(I_X, Z)$ holds for all Z that cause Y over a directed path that does not go through X in any expansion $S' = \langle V', E', DEP' \rangle$ of S.

The stochastic alternative version of Woodward's (2003) interventionist theory of causation can get on with the following notion of a stochastic i-expansion, which is much weaker than its counterparts for deterministic interventions. Like the notion of a weak i-expansion, stochastic i-expansions do only require an intervention variable for the target variable X w.r.t. the test variable Y; additional interventions for fixing other variables' values are not required. It is weaker than the notion of a weak i-expansion insofar as it only requires a stochastic intervention variable for X w.r.t. Y instead of a deterministic one:

Definition 5.16 (stochastic i-expansion) If $S = \langle V, E, DEP \rangle$ is a causal dependence model, then $S' = \langle V', E', DEP' \rangle$ is a stochastic i-expansion of S for target variable X w.r.t. test variable Y if and only if

(a) $\langle V', E' \rangle$ is an expansion of $\langle V, E \rangle$, and
(b) $V' \backslash V$ contains a stochastic intervention variable I_X for X w.r.t. Y (and nothing else), and
(c) for all $Z_i, Z_j \in V$ and all $M \subseteq V \backslash \{Z_i, Z_j\}$: $(IN)DEP(Z_i, Z_j | M)$ if and only if $(IN)DEP'(Z_i, Z_j | M, I_X = OFF)$, and
(d) for every $x \in val(X)$ there is an ON-value of the corresponding stochastic intervention variable I_X for X w.r.t. Y such that $DEP'(x, I_X = ON)$.

With help of the stochastic versions of the notions of an intervention variable (Definition 5.15) and a stochastic i-expansion (Definition 5.16) above, we can now define the following versions of Woodward's (2003) three kinds of causal relations. For total causation, the most reasonable modification discussed in Sect. 5.3 was Modification 5.2. Together with accepting stochastic intervention variables, this amounts to the following:

Definition 5.17 (total causation^S) If there exist stochastic i-expansions $S' = \langle V', E', DEP' \rangle$ of $S = \langle V, E, DEP \rangle$ for X w.r.t. Y, then X is a total cause of Y in S if and only if $DEP'(Y, I_X = ON)$ holds in some stochastic i-expansions S', where I_X is a stochastic intervention variable for X w.r.t. Y in S'.

The most reasonable modification discussed for direct causation was Modification 5.4; the most reasonable modification for contributing causation was

Modification 5.5. Together with accepting weak stochastic interventions, accepting these modifications amounts to the following two notions:

Definition 5.18 (direct causation^S) If there exist stochastic i-expansions $S' = \langle V', E', DEP' \rangle$ of $S = \langle V, E, DEP \rangle$ for X w.r.t. Y, then X is a direct cause of Y in S if and only if $DEP'(Y, I_X = ON | Par(Y) \backslash \{X\})$ holds in some stochastic i-expansions, where I_X is a stochastic intervention variable for X w.r.t. Y in S'.

Definition 5.19 (contributing causation^S) If every path between X and Y in $S = \langle V, E, DEP \rangle$ can be isolated and there exist stochastic i-expansions $S' = \langle V', E', DEP' \rangle$ of S for X w.r.t. Y, then X is a contributing cause of Y in S if and only if

(a) there is a chain π of direct causal relations (in the sense of Definition 5.18) from X to Y in S, and
(b) $DEP'(Y, I_X = ON | M)$ holds in some stochastic i-expansions S', where M blocks all causal paths connecting X and Y different from π.

Next I demonstrate that the three stochastic versions of an alternative interventionist theory above pick out exactly those directed causal chains $X \longrightarrow\!\!\!\longrightarrow Y$ they should, given the systems fulfill the respective requirements discussed in Sect. 5.2.2.

Total causation should coincide with the graph theoretical $X \longrightarrow\!\!\!\longrightarrow Y$ in systems which satisfy CMC and CFC:

Theorem 5.5 *If for all $X, Y \in V$ (with $X \neq Y$) there exist stochastic i-expansions $S' = \langle V', E', DEP' \rangle$ of $S = \langle V, E, DEP \rangle$ for X w.r.t. Y satisfying CMC and CFC, then for all $X, Y \in V$ (with $X \neq Y$): $X \longrightarrow\!\!\!\longrightarrow Y$ in S if and only if $DEP'(Y, I_X = ON)$ holds in some stochastic i-expansions S', where I_X is a stochastic intervention variable for X w.r.t. Y in S'.*

Proof Suppose $S = \langle V, E, DEP \rangle$ is a causal dependence model and for every $X, Y \in V$ (with $X \neq Y$) there are stochastic i-expansions of S for X w.r.t. Y satisfying CMC and CFC. Let X and Y be arbitrarily chosen elements of V (with $X \neq Y$). Let I_X be a stochastic intervention variable for X w.r.t. Y in S', where S' is one of the above presupposed stochastic i-expansions.

The if-direction: Suppose $DEP'(Y, I_X = ON)$ holds in S'. Then the d-connection condition implies that there is a causal path π d-connecting I_X and Y. Because of Definition 5.15(a), this path π cannot have the form $I_X \longleftarrow \ldots - Y$. Hence, π must have the form $I_X \longrightarrow \ldots - Y$. Since π d-connects I_X and Y (given $\emptyset$), π cannot feature colliders. Hence, π must be a directed path $I_X \longrightarrow\!\!\!\longrightarrow Y$. Now there are two possibilities: Either (A) π goes through X, or (B) π does not go through X. (B) is excluded by Definition 5.15(c). Therefore, (A) must be the case and π has to be a directed path $I_X \longrightarrow\!\!\!\longrightarrow X \longrightarrow\!\!\!\longrightarrow Y$. Since S' is a stochastic i-expansion of S, there must also be a corresponding path $X \longrightarrow\!\!\!\longrightarrow Y$ in S.

The only-if-direction: Suppose $X \longrightarrow\!\!\!\longrightarrow Y$ in S. Then $X \longrightarrow\!\!\!\longrightarrow Y$ is also in S'. Since I_X is an intervention variable for X, there is also a causal path $I_X \longrightarrow X$ in S'. The concatenation of paths $I_X \longrightarrow X$ and $X \longrightarrow\!\!\!\longrightarrow Y$ is $\pi : I_X \longrightarrow X \longrightarrow\!\!\!\longrightarrow Y$.

With faithfulness now $DEP'(Y, I_X)$ follows. Since $I_X = OFF$ is not correlated with any variable in V, Y must be dependent on some ON-value of I_X, i.e., $DEP(Y, I_X = ON)$. □

Next I show that our stochastic interventionist version of direct causation determines X to be a direct cause of Y in systems satisfying CMC and Min if and only if X is a causal parent of Y. The proof of Theorem 5.6 is based on the proof of theorem 3 in (Gebharter and Schurz 2014, sec. 6) adapted for dependence models.

Theorem 5.6 *If for all $X, Y \in V$ (with $X \neq Y$) there exist stochastic i-expansions $S' = \langle V', E', DEP' \rangle$ of $S = \langle V, E, DEP \rangle$ for X w.r.t. Y satisfying CMC and Min, then for all $X, Y \in V$ (with $X \neq Y$): $X \longrightarrow Y$ in S if and only if $DEP'(Y, I_X = ON | Par(Y) \backslash \{X\})$ holds in some stochastic i-expansions S', where I_X is a stochastic intervention variable for X w.r.t. Y in S'.*

Proof Suppose $S = \langle V, E, DEP \rangle$ is a causal dependence model and for every $X, Y \in V$ (with $X \neq Y$) there is a stochastic i-expansion of S for X w.r.t. Y satisfying CMC and Min. Now assume X and Y to be arbitrarily chosen variables in V with $X \neq Y$. Let I_X be a stochastic intervention variable for X w.r.t. Y in S' (where S' is one of the presupposed stochastic i-expansions of S w.r.t. Y).

The if-direction: Suppose $DEP'(Y, I_X = ON | Par(Y) \backslash \{X\})$ holds. Then, due to the d-connection condition, I_X and Y must be d-connected given $Par(X) \backslash \{Y\}$ over a causal path π. Because of Definition 5.15(a), π cannot have the form $I_X \longleftarrow \ldots - Y$. Hence, π must have the form $I_X \longrightarrow \ldots - Y$. Now either (A) π goes through X, or (B) π does not go through X.

Suppose (B) is the case and π does not go through X. Then, due to Definition 5.15(c), π cannot be a directed path $I_X \longrightarrow \longrightarrow Y$. Thus, either (i) π must have the form $I_X \longrightarrow \ldots - C \longrightarrow Y$ (with a collider on π), or (ii) π must have the form $I_X \longrightarrow \ldots - C \longleftarrow Y$. Assume (i) is the case. Then, since $C \neq X$, C must be an element of $Par(Y) \backslash \{X\}$. But then π would be blocked by $Par(Y) \backslash \{X\}$. Thus, (ii) must be the case and π must have the form $I_X \longrightarrow \ldots - C \longleftarrow Y$. But then there has to be a collider C^* on π that either is C or that is an effect of C. In both cases C^* is an effect of Y. But now I_X and Y can only be d-connected over π given $Par(Y) \backslash \{X\}$ if C^* is in $Par(Y) \backslash \{X\}$ or has an effect in $Par(Y) \backslash \{X\}$. But in both cases we would get a causal cycle $Y \longrightarrow \longrightarrow Y$, what is excluded by assumption. Thus, (A) must be the case.

Assume (A) is the case, meaning that π has the form $I_X \longrightarrow \ldots - X - \ldots - Y$. If π would have the form $I_X \longrightarrow \ldots - X - \ldots - C \longleftarrow Y$ (where C and X may be identical), then there has to be a collider C^* on π that is an effect of Y. π can only d-connect I_X and Y given $Par(Y) \backslash \{X\}$ if $Par(Y) \backslash \{X\}$ activates π. Thus, C^* has to be in $Par(Y) \backslash \{X\}$ or has to have an effect in $Par(Y) \backslash \{X\}$. In both cases we would end up with a causal cycle $Y \longrightarrow \longrightarrow Y$, what contradicts the assumption of acyclicity. Thus, π must have the form $I_X \longrightarrow \ldots - X - \ldots - C \longrightarrow Y$ (where C and X may be identical). Now either (i) $C = X$ or (ii) $C \neq X$. If (ii) would be the case, then C would be in $Par(Y) \backslash \{X\}$. Hence, $Par(Y) \backslash \{X\}$ would block π. Hence, (i) must be the case. Then π has the form $I_X \longrightarrow \ldots - X \longrightarrow Y$ and it follows from S' being a stochastic i-expansion of S that $X \longrightarrow Y$ in S.

The only-if-direction: Suppose $X \longrightarrow Y$ in S. Since S' is a stochastic i-expansion of S, $X \longrightarrow Y$ in S'. Since $X \longrightarrow Y$ in S' and Min is equivalent to Prod, it follows that $DEP'(y, x | Par(Y)\backslash\{X\} = r)$ holds for some $x \in val(X)$, for some $y \in val(Y)$, and for some instantiations r of $Par(Y)\backslash\{X\}$. Now suppose ON to be one of I_X's ON-values that is correlated with x. (Such an ON-value must exist due to Definition 5.16.) Then we have $DEP'(I_X = ON, x | r) \wedge DEP'(x, y | r)$.

Next we show that $DEP'(I_X = ON, y | r)$ holds in some stochastic i-expansion S' of S of the kind described above. To this end, let P be a probability distribution over V' represented by DEP'. Let $P^*(-)$ be defined as $P(- | r)$. We proceed by stating the following two probabilistically valid equations:

$$P^*(y | I_X = ON) = \sum_x P^*(y | x, I_X = ON) \cdot P^*(x | I_X = ON) \tag{5.11}$$

$$P^*(y) = \sum_x P^*(y | x) \cdot P^*(x) \tag{5.12}$$

Since X blocks path $I_X \longrightarrow X \longrightarrow Y$, we get from Equation 5.11 with the d-connection condition:

$$P^*(y | I_X = ON) = \sum_x P^*(y | x) \cdot P^*(x | I_X = ON) \tag{5.13}$$

Now there are two possibilities: Either (A) $P^*(y) \neq P^*(y | I_X = ON)$ holds, or (B) $P^*(y) = P^*(y | I_X = ON)$ holds. If (A) is the case, then we can infer $DEP'(I_X = ON, y | r)$.

Let us assume that (B) is the case. We already know that some x depends on $I_X = ON$ (in context r). Thus, $P^*(x) \neq P^*(x | I_X = ON)$ holds for some x. Let us refer to the set of all $x \in val(X)$ satisfying $P^*(x) \neq P^*(x | I_X = ON)$ by $M(X)$. Now there are two possibilities for $P^*(y)$ to equal $P^*(y | I_X = ON)$: Either (i) the differences in the weights $P^*(x)$ and $P^*(x | I_X = ON)$ in the sums in Equations 5.12 and 5.13, respectively, are exactly canceled by the differences of the conditional probabilities $P^*(y | x)$, or (ii) the conditional probabilities $P^*(y | x)$ are 0 for all $x \in M(X)$. Both cases constitute a case of non-faithfulness due to intransitivity (cf. Sect. 3.4.2). But recall that this specific kind of non-faithfulness arises due to a parameter fine-tuning. So if we can change the conditional probabilities $P^*(y | x)$, we will find a dependence between y and $I_X = ON$ (in context r). So if (B) is the case, we can choose another stochastic i-expansion S' of S whose associated dependence model DEP' represents probability distributions with slightly different parameters. Thus, $DEP'(I_X = ON, y | r)$ will hold in some stochastic i-expansion S' of S. $\square$

Last but not least, I show that our stochastic version of contributing causation (Definition 5.19) picks out all those directed causal connections $X \longrightarrow\longrightarrow Y$ that are productive. This result does only hold for CMC-satisfying causal systems in which every path connecting two variables can be isolated. Recall from Sect. 5.2.2 that without this restriction it cannot be guaranteed that productive paths can be identified.

Theorem 5.7 *If for all $X, Y \in V$ (with $X \neq Y$) there exist CMC-satisfying stochastic i-expansions $S' = \langle V', E', DEP' \rangle$ of $S = \langle V, E, DEP \rangle$ for X w.r.t. Y in which every directed causal path can be isolated and in which for every directed causal path there is a productive directed path, then for all $X, Y \in V$ (with $X \neq Y$): There is a path $\pi : X \longrightarrow\longrightarrow Y$ in S if and only if*

(a) *there is a chain π of direct causal relations (in the sense of Definition 5.18) from X to Y in S, and*
(b) *$DEP'(Y, I_X = ON | M)$ holds in some stochastic i-expansions S', where I_X is a stochastic intervention variable for X w.r.t. Y in S' and M is a subset of $V' \backslash \{I_X, X, Y\}$ that blocks all paths between I_X and Y different from π.*

Proof Suppose $S = \langle V, E, DEP \rangle$ is a causal dependence model and for every pair $\langle X, Y \rangle$ with $X, Y \in V$ (and $X \neq Y$) there exist stochastic i-expansions $S' = \langle V', E', DEP' \rangle$ of S for X w.r.t. Y satisfying CMC. Assume further that every directed path in S' can be isolated (where S' is one of the above presupposed stochastic i-expansions). Let X and Y be arbitrarily chosen elements of V (such that $X \neq Y$). Let I_X be a stochastic intervention variable for X w.r.t. Y in S'.

The if-direction: The if-direction is trivial.

The only-if-direction: Suppose there is a causal path $X \longrightarrow\longrightarrow Y$ in S. Since I_X is an intervention variable for X, there will be a causal path $\pi^* : I_X \longrightarrow X \longrightarrow\longrightarrow Y$ in S' with π as a subpath. Hence and because of our assumption above, there will also be a path $\pi^{**} : I_X \longrightarrow X \longrightarrow\longrightarrow Y$ that is productive. By assumption, there will also be a set $M \subseteq V' \backslash \{I_X, X, Y\}$ blocking all paths between I_X and Y different from π^{**} such that $DEP'(Y, I_X = ON | M)$. Since every arrow of path π^{**} is productive, there will also be a corresponding causal chain of directed causal relations (in the sense of Definition 5.18) from X to Y. $\qquad\square$

5.5 Conclusion

In this chapter I investigated the connection of Woodward's (2003) interventionist theory of causation to the theory of causal nets. After introducing Woodward's core definitions in Sect. 5.2.1, I gave a reconstruction of the Woodwardian core notions in terms of the theory of causal nets in Sect. 5.2.2. To connect causation to possible interventions it turned out as insufficient to represent dependencies among variables by a single probability distribution P. We used dependence models DEP instead. Dependence models allow for a connection between causal claims and counterfactual claims about what would happen under physically possible interventions, even if no such interventions ever occur in the actual world. In addition to Woodward's notions of an intervention variable and of total, direct, and contributing causation, I also introduced the notions of two different kinds of intervention expansions. These latter notions provide constraints on how a causal system of interest is allowed to be expanded when adding intervention variables for accounting for causal relationships among variables of the original system Woodward style.

One interesting discovery made in Sect. 5.2.2 was that Woodward's (2003) three causal notions are not expected to coincide with the corresponding graph theoretical causal connections ($X \longrightarrow Y$ in case X is a direct cause of Y and $X \longrightarrow\longrightarrow Y$ in case of total and contributing causation). It turned out that total causation should coincide with causal relevance ($X \longrightarrow\longrightarrow Y$) in systems satisfying CMC and CFC and that direct causation should coincide with $X \longrightarrow Y$ in systems satisfying CMC and Min. The existence of a causal chain $X \longrightarrow\longrightarrow Y$ should coincide with a relationship of contributing causation in case X and Y are connected by a productive chain $X \longrightarrow\longrightarrow Y$. But, as we saw, productiveness of directed causal paths is not determinable in every causal system. The problem is that there are causal systems in which directed causal paths cannot be isolated and individually tested for productiveness. So we contented ourselves with stating that contributing causal relationships should coincide with the existence of productive causal chains $X \longrightarrow\longrightarrow Y$ in CMC-satisfying causal systems in which every directed causal path can be isolated, meaning that all paths between X and Y different from this path can be blocked by conditionalizing on some subset M of the system of interest's variable set V.

In Sect. 5.3 I highlighted three kinds of problematic scenarios for Woodward's (2003) interventionist theory of causation. The theory does not give us the results it should when the sets of variables we are interested in contain variables one cannot manipulate in the strong Woodwardian sense (Sect. 5.3.1). I discussed several possible solutions to this problem. It turned out that the most reasonable solution is to replace Woodward's explicit definitions of total, direct, and contributing causation by partial definitions whose if-conditions require the existence of suitable intervention variables. This has the effect that the modified interventionist theory, instead of producing false results, is no longer applicable to the problematic systems—it keeps silent about causation in such systems. To explicate the meaning of causation, one could supplement the modified interventionist theory by other principles, e.g., by the causal Markov condition and the causal minimality condition, which are assumed to hold also in systems containing variables one cannot intervene on in the strong deterministic sense. Another possibility would be to allow for stochastic interventions. In that case one has to replace Woodward's strong fixability condition implemented in his definitions of direct and contributing causation by a weaker conditionalization condition: For direct causation between X and Y a dependence between I_X and the test variable Y conditional on $Par(Y)\backslash\{X\}$ is necessary and sufficient (given CMC and Min are satisfied). For contributing causation between X and Y a causal path $\pi : X \longrightarrow\longrightarrow Y$ and a dependence between I_X and test variable Y conditional on a set M that isolates π is necessary and sufficient (in

CMC-satisfying causal systems in which for every directed causal path there is at least one productive directed causal path).

In Sect. 5.3.2 we discovered that Woodward's (2003) notion of an intervention variable does not what it was ultimately intended for. It does not pick out those variables as intervention variables which allow correct inference about whether X is a cause* of Y. The problem arises because the notion of an intervention variable allows for intermediate causes to turn out as intervention variables for their effects w.r.t. their causes. In that case, a dependence between the intervention and its cause implies that the intervention's effect causes its cause. The problem can be solved by requiring intervention variables to be exogenous. Another problem (Sect. 5.3.3) was that the effect Y of a (physically or nomologically) deterministic cause X cannot be fixed by interventions (at least not by interventions which cause Y over a path not going through X). This leads to false results in deterministic causal chains and deterministic common cause structures. The problem can be solved quite analogously to the first problem discussed in Sect. 5.3.1. One can replace Woodward's explicit definitions by partial definitions whose if-conditions require suitable intervention variables. In that case, the modified theory keeps silent about causation in systems featuring deterministic causal chains or deterministic common cause paths. Alternatively, one can allow for stochastic interventions and replace Woodward's original fixability condition by a weaker fixability condition based on conditionalization instead of interventions.

In Sect. 5.4 I first proposed an alternative version of Woodward's (2003) interventionist theory of causation. This alternative theory is the result of accepting the most reasonable modifications discussed in Sect. 5.3 for deterministic interventions. It can avoid the problems observed in Sect. 5.3. In addition, I showed that within this new theory total causation coincides with causal relevance ($X \longrightarrow \longrightarrow Y$) in systems satisfying CMC and CFC, that direct causation coincides with $X \longrightarrow Y$ in systems satisfying CMC and Min, and that contributing causation coincides with productive directed causal paths in CMC-satisfying systems in which every directed path π's alternative paths can be blocked by some set M. In Sect. 5.4.2 I did the same thing for the stochastic variant of the alternative to Woodward's theory which resulted from the most reasonable modifications for stochastic interventions discussed in Sect. 5.3.

What this part of the book was intended to show is, among other things, that intuitively nice theories may not always be the best choice. Woodward's (2003) interventionist theory comes with several problems which could be easily avoided when one endorses the technically more complex theory of causal Bayes nets instead. In addition, it seems to be the case that whenever Woodward's interventionist theory can be successfully applied, the same results can be obtained by using the theory of causal Bayes nets. I would like to point everyone interested in how the two theories can handle more philosophical issues related to causation to a recent debate within the philosophy of mind: Several authors (e.g., Raatikainen 2010; Shapiro 2010; Shapiro and Sober 2007; Woodward 2008a) have recently argued that one can escape causal exclusion arguments when subscribing to Woodward's interventionist theory of causation. As Baumgartner (2009, 2010) has convincingly shown, however, Woodward's original interventionist framework gives, if one takes a closer look to its core definitions, actually rise to a much stronger version of causal exclusion argument (requiring less or weaker premises). Woodward (2015) has meanswhile responded to Baumgartner's results by coming up with a modified interventionist theory of causation which—at least in principle—allows for mental causation. However, just as done with the problems of Woodward's original theory highlighted in Chap. 5, one can use the theory of causal Bayes nets as a diagnostic tool for assessing whether Woodward's modified interventionist theory can actually deliver support for mental causation. It turns out that the theory has a built in blind spot when it comes to testing for causal efficacy of mental variables supervening on physical variables. And, even more surprisingly, it turns out that it is not even logically possible to intervene on such mental variables (Gebharter 2015). So the whole project of providing empirical support within an interventionist framework for mental causation was somehow deemed to failure from the beginning. To get these results, however, a much richer and technically more sophisticated framework is required than Woodward's interventionism.

Chapter 6
Causal Nets and Mechanisms

Abstract In this chapter I enter the new mechanist debate within the philosophy of science. I discuss a proposal how to model mechanisms made by Casini, Illari, Russo, and Williamson and present three problems with their approach. I then present my alternative approach of how to represent mechanisms, compare it with Casini et al.'s approach, and discuss Casini's recent objections to my approach. I also make a suggestion how constitutive relevance relations could be represented within my approach. In the rest of this chapter I extend the approach in such a way that it can also account for the diachronic character of mechanisms and the fact that many mechanisms feature causal cycles.

6.1 Introduction

In this chapter I try to apply the theory of causal nets to mechanisms as they are discussed in nowadays philosophy of science literature. The new mechanist movement started with papers such as Bechtel and Richardson (2000) and Machamer et al. (2000). Mechanists claim that many (or even all) explanations in the sciences are best understood as mechanistic explanations, i.e., explanations of the phenomena of interest by pointing at these phenomena's underlying mechanisms. Illari and Williamson (2012, p. 119) cite the following three characterizations of mechanisms as the most important ones:

> Mechanisms are entities and activities organized such that they are productive of regular changes from start or set-up to finish or termination conditions. (Machamer et al. 2000, p. 3)

> A mechanism for a behavior is a complex system that produces that behavior by the interaction of a number of parts, where the interactions between parts can be characterized by direct, invariant, change-relating generalizations. (Glennan 2002, p. S344)

> A mechanism is a structure performing a function in virtue of its component parts, component operations, and their organization. The orchestrated functioning of the mechanism is responsible for one or more phenomena. (Bechtel and Abrahamsen 2005, p. 423)

© Springer International Publishing AG 2017   115
A. Gebharter, *Causal Nets, Interventionism, and Mechanisms*,
Synthese Library 381, DOI 10.1007/978-3-319-49908-6_6

Other important characterizations are given by Craver (2007b) and Woodward (2002, p. S375). (See also Glennan's 1996, p. 52 older but still frequently cited definition of a mechanism.) The most general characterization of a mechanism is probably Illari and Williamson's:

> A mechanism for a phenomenon consists of entities and activities organized in such a way that they are responsible for the phenomenon. (Illari and Williamson 2012, p. 120)

There are a lot of controversies and ambiguities in the discussion of the merits and shortcomings of the diverse definitions of a mechanism as well as in the discussion about the ontological status of mechanisms, whether they require two kinds of entities (which Machamer et al. 2000 call entities and activities) or just one kind of entity, *viz.* entities that are causally interacting, as, for example, Glennan (2002) claims (see also Glennan 2009). It is also not clear whether the mechanistic framework presupposes some kind of reductionism (cf. Eronen 2011, 2012). In this chapter I try to bracket all of these controversies and focus on what is commonly accepted.

One thing that is commonly accepted among mechanists is that mechanisms are systems that exhibit certain input-output behaviors. They are wholes consisting of several parts (in a mereological sense). These parts causally interact with each other and it is this causal interaction that somehow brings about the phenomenon (input-output behavior) of interest. Mechanisms are typically spatio-temporal systems in which the causal interaction of parts requires some time. In addition, mechanisms are oftentimes self-regulating systems which may include a lot of causal loops or causal feedback. As a last feature, mechanisms are typically hierarchically organized, i.e., a mechanism may consist of several submechanisms, which may themselves again consist of submechanisms, and so on.

My claim is that mechanisms can be modeled nicely by means of causal nets. Such a formal representation of mechanisms has the advantage over a qualitative description of mechanisms, which is what we mostly find in the philosophical literature on mechanisms, that it allows for probabilistic and quantitative reasoning, prediction, and explanation.[1] Let me illustrate this by means of the following example: Assume we want to predict the probability of a certain acceleration a of a car when the car's gas pedal is pressed with pressure p. Characterizations of mechanisms like the ones given above may qualitatively tell us how the gas pedal is connected to the car's wheels, its gas tank, its motor, etc., but they do not provide any detailed information about values p and a. Though qualitative mechanistic descriptions may be good for understanding the structure of a mechanism, they are not appropriate for quantitative probabilistic prediction.[2] Answering questions about how p relates to a requires a formalism capable of computing the numerical effects of certain values p on specific values a.

---

[1]I assume here that all of the relevant causal and probabilistic information is already available.

[2]For a comparison of qualitative descriptions of mechanisms versus quantitative models of mechanisms by means of causal nets see, for example, Gebharter and Kaiser (2014).

Such a formalism does not only have to allow for computing a (or the probability of a) given p, it does also have to meet at least some of the characteristic features of mechanisms described above. As far as I know, there are only two modeling approaches for mechanisms by means of causal nets. Both focus on the hierarchical organization of mechanisms and not so much on their other characteristic marks. One of these accounts is developed in detail in Casini et al. (2011), the other one in Gebharter (2014). Casini et al. suggest to model mechanisms by so-called recursive Bayesian networks (RBNs), which were developed by Williamson and Gabbay (2005) as tools for modeling nested causal relationships. They suggest to represent a mechanism as a special kind of variable in a recursive Bayesian network.

This chapter brings together the results obtained in Gebharter (2014, 2016) and Gebharter and Schurz (2016) and presents them in much more detail. I will also defend my approach against recent objections made by Casini (2016) and make a novel proposal how constitutive relevance relations (if one wants to have them) can be represented and maybe discovered within my approach. I will start by presenting the recursive Bayesian network formalism and Casini et al.'s (2011) approach for modeling mechanisms in Sect. 6.2. In Sect. 6.3 I will highlight three problems with the RBN approach. The first problem is that Casini et al.'s approach does not feature a graphical representation of how a mechanism's micro structure is causally connected to a mechanism's input and output variables at the macro level. Such a representation is important for mechanistically explaining how certain inputs to a mechanism (such as p in the car example given above) produce certain outputs (such as a). The second problem is that the RBN approach leads to the consequence that a mechanism's macro behavior cannot be changed by manipulating the micro behavior of its parts by means of intervention variables. This contradicts scientific practice. According to Craver (2007a,b) biologists carry out so-called top-down and bottom-up experiments to pick out the micro parts of the mechanism constitutively relevant for the mechanism's macro behavior. Such experiments should be representable within the model. The third problem is that also the method Casini et al. suggest for computing the effects of manipulations without representing these manipulations by means of intervention variables is inadequate. It regularly leads to false results concerning the probabilities of higher level inputs as well as the probabilities of same-level causes of the variables one manipulates. Meanwhile, Casini (2016) has responded to the first two problems. I will also discuss Casini's responses in detail in Sect. 6.3.

In Sect. 6.4 I will present the alternative modeling approach developed in (Gebharter 2014). This alternative approach does not fall victim to the three problems mentioned. In this approach, mechanisms will not be represented by variables, but by certain subgraphs of causal models possibly featuring bi-directed causal arrows ('$\longleftrightarrow$') in addition to the ordinary causal arrows ('$\longrightarrow$'). This alternative modeling approach does not only avoid the problems of the RBN approach, but also suggests to develop new procedures for discovering submechanisms, i.e., the causal structure inside the arrows. (For an example, see Murray-Watters and Glymour 2015.) It also allows for applying results from the statistics and machine learning literature. Zhang (2008), for example, develops a method for computing the effects of interventions in models featuring bi-directed arrows (see also Tian and Pearl 2002). Richardson (2009) presents a factorization criterion for graphs with bi-directed edges which is equivalent to the d-connection condition (adapted for such graphs). I proceed by discussing some of the philosophical consequences of the alternative approach for modeling mechanisms to be developed.

If my representation of mechanisms is adequate, then it seems that constitutive relevance relations in the sense of Craver (2007a,b) are maybe not that special and important for explanation and prediction as many mechanists think. In Sect. 6.5 I argue that constitutive relevance shares the same formal properties with direct causal relations which are captured by the causal Markov condition: They produce the Markov factorization. They also produce the same screening off and linking-up phenomena as direct causal relations. This means that constitutive relevance can be represented like a special kind of causal relation in MLCMs. Because of this, standard search procedures for causal structure can—so it seems—also be used to uncover constitutive relevance relations in mechanisms. I demonstrate how such a standard search procedure for causal connections could be used to uncover constitutive relevance relations by means of a simple toy example. Finally, I discuss the empirical adequacy of my proposal of how to model constitutive relevance in MLCMs and also some philosophical consequences of this approach.

In the last section of Chap. 6, *viz.* Sect. 6.6, I focus on two of the other characteristic marks of mechanisms mentioned earlier: the diachronic aspect of mechanisms, i.e., the fact that mechanisms typically produce their target phenomenon over a certain period of time, and the fact that many mechanisms feature feedback. In a follow up paper to Casini et al. (2011), Clarke et al. (2014) expand Casini et al.'s RBN approach for representing mechanisms in such a way that it can also be used to model causal feedback and the diachronic aspect of mechanisms. Clarke et al. use dynamic Bayesian networks (cf. Murphy 2002) for the latter. I use cyclic and dynamic causal models also allowing for bi-directed arrows and show that the MLCM approach can also represent mechanisms involving causal feedback. I finally compare my representation method for such mechanisms with Clarke et al.'s representation and highlight several advantages and open problems of my approach.

6.2 The RBN Approach

6.2.1 *Recursive Bayesian Networks*

In this subsection I will introduce the basics of Williamson and Gabbay's (2005) *recursive Bayesian network* (RBN) formalism, which was initially developed for modeling nested causal relationships. I will just present the formalism in this subsection, without saying anything about nested causation. The reason for this is that nested causal relationships will not be of any relevance in the remainder of Chap. 6.

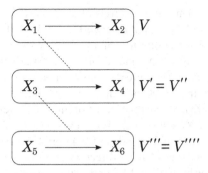

Fig. 6.1 Structure of an RBN. X_1 and X_3 are network variables, while all other variables in the diagram are simple variables. The *dashed lines* indicate which BNs are possible values of which variables. The BNs $S' = \langle V', E', P' \rangle$ and $S'' = \langle V'', E'', P'' \rangle$, for example, are the possible values of the network variable X_1

An RBN is a special kind of Bayesian network, *viz.* a Bayesian network whose variable set contains some variables whose values are, again, Bayesian networks. Such vertices whose values are again BNs are called *network variables*, while Williamson and Gabbay (2005) call variables without BNs as values *simple variables*.

Let me illustrate some new notational conventions which will be useful when working with RBNs by means of the following example. (This exemplary RBN's structure is depicted in Fig. 6.1.) Assume $S = \langle V, E, P \rangle$ is an RBN with $V = \{X_1, X_2\}$ and topological structure $X_1 \longrightarrow X_2$. Assume further that X_1 is a network variable with the two BNs $S' = \langle V', E', P' \rangle$ and $S'' = \langle V'', E'', P'' \rangle$ as possible values, while X_2 is a simple variable. Suppose $V' = V'' = \{X_3, X_4\}$ and that the topological structure of both BNs S' and S'' is $X_3 \longrightarrow X_4$. Let us assume, in addition, that X_3 is a network variable with the two BNs $S''' = \langle V''', E''', P''' \rangle$ and $S'''' = \langle V'''', E'''', P'''' \rangle$ as possible values, while X_4 is a simple variable. Again, V''' should be identical to V'''', which we assume to be identical to the set $\{X_5, X_6\}$, and S''' and S'''' shall share the same graph $X_5 \longrightarrow X_6$. This time, both X_5 and X_6 shall be simple variables.

Note that there is some idealization involved in the RBN depicted in Fig. 6.1. The RBN formalism in principle also allows for different causal structures of S' and S'' as well as of S''' and S''''. More generally, it also allows for $V' \neq V''$ and $V''' \neq V''''$. However, when used to model mechanisms, it will almost always hold that $V' = V''$ and $V''' = V''''$. Hence, V' and V'' as well as V''' and V'''' will almost always share the same causal structure. Hence, I will only have a look at RBN models meeting these idealizations in this chapter. But note that the problems the RBN approach has when it comes to modeling mechanisms I describe in subsequent sections generalize to RBN models of mechanisms not meeting these idealizations.

In RBNs, the family terminology introduced in Sect. 2.7 applies to all the BNs in the given ordering. X_1, for example, is a parent of X_2, which is X_1's child in BN S's graph $G = \langle V, E \rangle$, etc. But it will be of help to also have notions for relationships between variables located at different levels of an RBN. If Y is a variable of one of the BNs which are possible values of a network variable X, then Y is called a *direct inferior* of X. The set of direct inferiors of a variable X is denoted by '$DInf(X)$'. In our exemplary RBN in Fig. 6.1 X_3 and X_4 are direct inferiors of X_1, while X_5 and X_6 are direct inferiors of X_3. The (direct or indirect) *inferiors* of a variable X are the variables Y in $DInf(X)$, the variables which are direct inferiors of those variables Y (i.e., elements of $DInf(Y)$), and so on. The set of inferiors is referred to via '$Inf(X)$'. So X_3, X_4, X_5, and X_6 are X_1's inferiors, while X_5 and X_6 are X_3's inferiors. X is called a *superior* of Y if and only if Y stands in the inferiority relation to X, and X is a *direct superior* of Y if and only if Y is a direct inferior of X. In our exemplary RBN, thus, X_1 is a direct superior of X_3 and X_4, and a superior of X_3, X_4, X_5, and X_6, while X_3 is a superior (and also a direct superior) of X_5 and X_6. The set of direct superiors of a variable X is symbolized by '$DSup(X)$', the set of superiors by '$Sup(X)$'.

If $S = \langle V, E, P \rangle$ is an RBN, for example the RBN depicted in Fig. 6.1, then **V** shall be the RBN's variable set V under the transitive closure[3] of the inferiority relation. So **V** contains not only the variables of the RBN's top level (X_1 and X_2 in our example), but also all the variables of any lower level BN. In our exemplary RBN, **V** would be identical to $\{X_1, \ldots, X_6\}$. The bold '**N**' shall stand for the set of all variables in **V** that are network variables. In our example, **N** would, hence, be identical to the set $\{X_1, X_3\}$.

The standard Bayesian net formalism allows us to compute probabilities only within the individual BNs in an RBN's hierarchy. Each of these BNs has a (typically different) associated probability distribution $(P, P', \ldots, P'''')$ in our example). But one can also construct a probability distribution **P** over **V** that allows for probabilistic inference across the diverse levels of an RBN. This is one major (if not the) main selling point of RBNs. For the construction of this probability distribution **P** we need one additional notion: the notion of an RBN's flattening. Let **V** of our RBN be $\{X_1, \ldots, X_m\}$, and **N** be $\{X_{j_1}, \ldots, X_{j_k}\}$. Suppose $\mathbf{n} = \{x_{j_1}, \ldots, x_{j_k}\}$ to be an instantiation of the network variables $X_{j_1}, \ldots, X_{j_k}$ to their values $x_{j_1}, \ldots, x_{j_k}$. Then for every such possible instantiation $\mathbf{n} = \{x_{j_1}, \ldots, x_{j_k}\}$ we can construct a Bayesian network which is called the RBN's *flattening* w.r.t. $\mathbf{n}$ ('$\mathbf{n}\downarrow$' for short). A flattening's nodes are the simple variables in **V** plus the instantiation of network variables in $\mathbf{N} = \{X_{j_1}, \ldots, X_{j_k}\}$ to their values $x_{j_1}, \ldots, x_{j_k}$. The flattening's topology is constructed in the following way: Draw an arrow exiting X and pointing at Y if and only if X is either a parent or a direct superior of Y. As a last step, we have to determine the flattening $\mathbf{n}\downarrow$'s associated probability distribution $P^{\mathbf{n}\downarrow}$. This can be done by the following equation (where X_{j_l} are the direct superiors of X_i):

$$P^{\mathbf{n}\downarrow}(x_i|par(X_i), dsup(X_i)) = P_{x_{j_l}}(x_i|par(X_i)). \tag{6.1}$$

---

[3]A relation R's *transitive closure* R' is defined as $R' = \{\langle u, v \rangle : \exists w_1, \ldots \exists w_n (\langle u, w_1 \rangle \in R \wedge \ldots \wedge \langle w_n, v \rangle \in R)\}$.

Fig. 6.2 The flattening $\mathbf{n}{\downarrow}$
w.r.t. $\mathbf{n} = \{X_1 = S', X_3$
$= S'''\}$'s associated graph

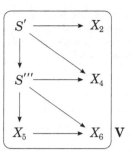

Let me now demonstrate on our exemplary RBN introduced above how one of its flattenings can be constructed according to the procedure described. Recall that in this RBN $\mathbf{V}$ is $\{X_1, \ldots, X_6\}$, while $\mathbf{N}$ is $\{X_1, X_3\}$. Now assume we are interested in the flattening $\mathbf{n}{\downarrow}$ determined by $\mathbf{n} = \{X_1 = S', X_3 = S'''\}$. According to the above instructions for constructing flattenings, the flattening's set of vertices is $\{X_1 = S', X_2, X_3 = S''', X_4, X_5, X_6\}$. To determine its topological structure, we have to draw an arrow from a vertex in this set to another one if and only if the former is either a parent or a direct superior of the latter. When we follow these instructions, the resulting graph of the flattening is the one depicted in Fig. 6.2. According to Eq. 6.1, the flattening $\mathbf{n}{\downarrow}$'s associated probability distribution $P^{\mathbf{n}{\downarrow}}$ is $P^{\mathbf{n}{\downarrow}}(X_1 = S') = 1$, $P^{\mathbf{n}{\downarrow}}(X_2|X_1) = P(X_2|X_1)$, $P^{\mathbf{n}{\downarrow}}(X_3 = S''') = 1$, $P^{\mathbf{n}{\downarrow}}(X_4|X_1 = S', X_3 = S''') = P'(X_4|X_3 = S''')$, $P^{\mathbf{n}{\downarrow}}(X_5|X_3 = S''') = P'''(X_5)$, $P^{\mathbf{n}{\downarrow}}(X_6|X_3 = S''', X_5) = P'''(X_6|X_5)$.

Assume we have constructed all the flattenings of the RBN of interest w.r.t. every possible instantiation of network variables in $\mathbf{N}$. Then we can, on the basis of these flattenings, construct the much sought-after probability distribution $\mathbf{P}$ over $\mathbf{V}$, which finally allows for probabilistic reasoning across the levels of the RBN, by means of the following formula (where the probabilities on the right hand side of the '=' are determined by the flattening $\mathbf{n}{\downarrow}$ induced by $x_1, \ldots, x_m$):

$$\mathbf{P}(x_1, \ldots, x_m) = \prod_{i=1}^{m} P^{\mathbf{n}{\downarrow}}(par(X_i), dsup(X_i)) \tag{6.2}$$

In case of our exemplary RBN introduced above, the resulting probability distribution $\mathbf{P}$ over this RBN's variable set $\mathbf{V}$ would be $\mathbf{P}(X_1, X_2, X_3, X_4, X_5, X_6) = P^{\mathbf{n}{\downarrow}}(X_1) \cdot P^{\mathbf{n}{\downarrow}}(X_2|X_1) \cdot P^{\mathbf{n}{\downarrow}}(X_3|X_1) \cdot P^{\mathbf{n}{\downarrow}}(X_4|X_1, X_3) \cdot P^{\mathbf{n}{\downarrow}}(X_5|X_3) \cdot P^{\mathbf{n}{\downarrow}}(X_6|X_3, X_5)$, where the probabilities on the right hand side of the '=' are, again, determined by the flattening $\mathbf{n}{\downarrow}$ induced by $X_1, \ldots, X_6$.

An RBN's probability distribution $\mathbf{P}$ over $\mathbf{V}$ determined by Eq. 6.2 determines again a Bayesian network over $\mathbf{V}$, which can be used to simplify probabilistic reasoning across the RBN's levels. In case of our exemplary RBN, this BN determined by $\mathbf{P}$ would have the graphical structure depicted in Fig. 6.3.

Fig. 6.3 The graph of the
BN determined by our
exemplary RBN's probability
distribution **P**

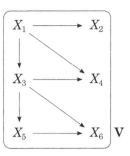

Just as BNs satisfy the Markov condition due to Markov factorizability (see
Sect. 2.7), RBNs satisfy the so-called *recursive Markov condition* (RMC) due
to Eq. 6.2. The formulation of RMC is quite analogously to MC's formulation
Definition 2.4:

Definition 6.1 (recursive Markov condition) An RBN $S = \langle V, E, P \rangle$ satisfies the
recursive Markov condition if and only if $INDEP_\mathbf{P}(X, NID(X)|DSup(X) \cup Par(X))$
holds for all $X \in \mathbf{V}$.

'$NID(X)$' refers to the *non-inferiors-or-descendants* of X, i.e., the set
$\mathbf{V}\backslash(Inf(X) \cup Des(X))$. One can interpret the horizontal arrows in the graph
depicted in Fig. 6.3 as indicating the direct ancestor/descendant relation, while
the vertical as well as the diagonal arrows can be interpreted as indicating the
direct superior/inferior relation. So RMC would imply for our exemplary RBN the
following probabilistic independencies (as well as all probabilistic independencies
implied by them):

- $INDEP_\mathbf{P}(X_2, \{X_3, X_4, X_5, X_6\}|\{X_1\})$
- $INDEP_\mathbf{P}(X_3, \{X_2\}|\{X_1\})$
- $INDEP_\mathbf{P}(X_4, \{X_2, X_5, X_6\}|\{X_1, X_3\})$
- $INDEP_\mathbf{P}(X_5, \{X_1, X_2, X_4\}|\{X_3\})$
- $INDEP_\mathbf{P}(X_6, \{X_1, X_2, X_4\}|\{X_3, X_5\})$

6.2.2 Modeling Mechanisms with RBNs

Let me now illustrate how Casini et al. (2011) suggest to use RBNs to model nested
mechanisms. For this purpose I want to introduce a very simple toy mechanism: the
water dispenser mechanism. The water dispenser is a device that consists of a button
(formally represented by B) and a water temperature regulation unit (modeled by a
variable D). B has two possible values, *viz.* 1 and 0, which indicate whether the
button is pressed or not pressed, respectively. When the water dispenser's button
is pressed ($B = 1$), then the machine produces a cup of water whose temperature
(represented by W) is close to the room temperature (modeled by T). If the button is
not pressed, on the other hand, the device dispenses cold water. The causal structure
regulating the behavior of the water dispenser is depicted in Fig. 6.4.

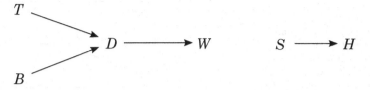

Fig. 6.4 The causal structure underlying the water dispenser device's behavior is depicted on the *left hand side*; the causal connection of the parts of the mechanism represented by D on the *right hand side*

In this example, the water regulation unit represented by D would be a mechanism with two parts: a temperature sensor (modeled by a variable S) and a heater (represented by H), whose causal relationship is also depicted in Fig. 6.4. Now Casini et al. (2011) suggest to represent a mechanism's possible states by an RBN's network variable. Each of the mechanism's states should be represented by a causally interpreted BN providing information about the causal connections between the mechanism's micro parts. Thus, we would have to model the water dispenser by means of an RBN $\langle V, E, P \rangle$, where $V = \{T, B, D, W\}$, while $\langle V, E, P \rangle$'s topological structure is the one depicted on the left hand side of Fig. 6.4. T, B, and W would be simple variables, while D would be a network variable with the two possible values BN_1 (a BN representing a temperature regulation unit that is working) and BN_0 (a BN modeling a temperature regulation unit that is not working). Note that BN_1 as well as BN_0 are both assumed to have the same set of vertices $\{S, H\}$ and also the same causal structure (depicted on the right hand side of Fig. 6.4). The only difference between BN_1 and BN_0 is due to their associated probability distributions. $D = BN_1$ describes the heater working on a level that corresponds to the input the temperature sensor S measures, and thus, the probability of high H values, for example, will be high when S's value is high, and low, when S's value is low, and so forth. If $D = BN_0$, then the dispenser's temperature regulation unit is not working at all. This will lead to a cup of cold water dispensed and to a probabilistic independence between S and H in BN_0.

One of the RBN approach's merits should be to allow for quantitative probabilistic reasoning across a mechanism's diverse levels. For this purpose, as described in Sect. 6.2.1, we have to compute the probability distribution **P** over **V** = $\{T, B, D, W, S, H\}$. This has to be done by means of the RBN's flattenings. In case of our exemplary mechanism, there are two flattenings. (One of them is depicted in Fig. 6.5.) The probability distribution of the RBN's flattening w.r.t. $D = BN_1$ can, according to Eq. 6.1, be determined to be $P^{BN_1\downarrow}(T) = P(T)$, $P^{BN_1\downarrow}(B) = P(B)$, $P^{BN_1\downarrow}(D = BN_1) = 1$, $P^{BN_1\downarrow}(S|D = BN_1) = P^{BN_1}(S)$, $P^{BN_1\downarrow}(H|D = BN_1, S) = P^{BN_1}(H|S)$, $P^{BN_1\downarrow}(W|D = BN_1) = P(W|D = BN_1)$. The RBN's flattening w.r.t. $D = BN_0$ has the following probability distribution associated: $P^{BN_0\downarrow}(T) = P(T)$, $P^{BN_0\downarrow}(B) = P(B)$, $P^{BN_0\downarrow}(D = BN_0) = 1$, $P^{BN_0\downarrow}(S|D = BN_0) = P^{BN_0}(S)$, $P^{BN_0\downarrow}(H|D = BN_0, S) = P^{BN_0}(H|S)$, $P^{BN_0\downarrow}(W|D = BN_0) = P(W|D = BN_0)$. By means of Eq. 6.2 we can now compute the RBN's probability distribution **P** over

Fig. 6.5 Graph of the flattening of the RBN modeling the water dispenser w.r.t. $D = BN_1$. The *continuous arrows represent* intra level direct causal dependencies; the *dashed arrows* can be interpreted as constitution relations in the sense of Craver (2007a,b)

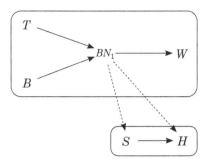

V (where the probabilities on the right hand side of the '=' are determined by the flattening $\mathbf{n}\downarrow$ induced by T, B, D, W, S, H): $\mathbf{P}(T, B, D, S, H, W) = P^{\mathbf{n}\downarrow}(T) \cdot P^{\mathbf{n}\downarrow}(B) \cdot P^{\mathbf{n}\downarrow}(D|T, B) \cdot P^{\mathbf{n}\downarrow}(S|D) \cdot P^{\mathbf{n}\downarrow}(H|D, S) \cdot P^{\mathbf{n}\downarrow}(W|D)$.

6.3   Three Problems with the RBN Approach

6.3.1   Problem (i): Missing Causal Information

The first problem I want to discuss is that the topology of an RBN used for modeling a mechanism does not feature the causal relationships of the mechanism's input and output at the macro level to its micro structure. Such information is, however, crucial when it comes to mechanistic explanation and also when it comes to questions about manipulation and control. Let me begin to explain the problem with mechanistic explanation by means of our toy example, i.e., the water dispenser. In this example, D represents a mechanism (the temperature regulation unit). T and B are D's input variables, and W is D's output variable. What should be mechanistically explained is the mechanism's behavior as an input-output function (cf., e.g., Craver 2007b, p. 145f.), i.e., a bunch of phenomena including the following ones:

- The water dispenser dispenses (with high probability) water close to the room temperature when the tempering button is pressed.
- The water dispenser dispenses (with high probability) cold water when the tempering button is not pressed.

What mechanistic explanation now is supposed to do is to tell a story about how certain inputs lead to certain outputs by describing how the causal influences of the inputs are propagated through the mechanism's micro structure. This requires information about how the input variables (T and B in our example) are causally connected to the mechanism's (i.e., D's) causal micro structure (represented by $S \longrightarrow H$), and how this micro structure, in turn, causally produces the mechanism's output (represented by W in case of the water dispenser). The RBN used to model the water dispenser device does not meet these requirements—it does not contain any arrows going from one of the input variables T or B to the micro variables S or H, nor any arrows going from S or H to the output variable W. The model does not tell us any causal story about how certain T- and B-values are causally connected through the causal micro structure of D to certain W-values. Note that this is only a problem when it comes to explanation, and not so much for prediction (based on observation; but see below). It may still be the case that the RBN's probability distribution $\mathbf{P}$ over $\mathbf{V}$ gives us the right probabilities.[4]

Let me illustrate the explanation problem by discussing a concrete exemplary phenomenon to be explained: One such phenomenon is that the room temperature (T) is screened off from the temperature of the water dispensed (W) if the tempering button is not pressed ($B = 0$). So let us ask *why* this is the case. The answer cannot be given by the RBN's probability distribution $\mathbf{P}$ alone, but does also have to cite some causal information: T is causally relevant for W because it first causes S, which then causes H, which is, in turn, directly causally relevant for W; B is causally relevant for W only over a directed causal path going (only) through H. When $B = 0$, this turns the heating unit H off, what means that $B = 0$ forces H to take a certain value, *viz. OFF*. Since the only directed causal path going from T to W goes through H, and $B = 0$ blocks this causal path, T and W become probabilistically independent in case $B = 0$. So what one would need to explain this input-output behavior is a causal structure like the one depicted in Fig. 6.6, which is not provided by the water dispenser's RBN.

That RBNs do not feature the causal relationships of a mechanism's input and output variables to its causal micro structure does not only lead to problems when it comes to mechanistic explanation, but also when it comes to questions concerning manipulation and control. A causal structure, like the one depicted in Fig. 6.6, for example, provides information about how certain causes' influences on certain variables of interest can be modified by manipulation. If we want to increase T's causal effect on W, for instance, this causal structure tells us that we can do so by boosting T's influence on the temperature sensor S or by amplifying S's influence on the heater H.[5] Such manipulations could, for example, be carried out by putting a heat source to the sensor modeled by S. To decrease T's effect on W, one could,

Fig. 6.6 Causal information needed to mechanistically explain why T and W become probabilistically independent when not pressing the tempering button ($B = 0$)

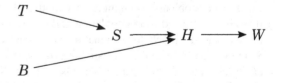

---

[4]But note that probabilities are typically not enough for explanation. Explaining an event e_2 by means of an event e_1 requires information about the causal pathways which lead from e_1 to e_2 (cf. Salmon 1984; Woodward 2011b, sec. 4).

[5]Note that the intervention variables needed for amplifying (and also for decreasing) the influence of a certain variable have to be stochastic interventions (cf. Eberhardt and Scheines 2007, sec. 2).

for example, put an ice cube to S. The water dispenser's RBN, on the other hand, does not provide such information: It does not provide information about how S is causally connected to T and W; all the information we have about how S-changes influence T and W is provided by the probability distribution $\mathbf{P}$ over $\mathbf{V}$.

I have first presented the problem explained in this subsection in Gebharter (2014). Meanwhile, Casini (2016) responded to my criticism. I will briefly discuss his objection here. Casini's response to the first problem discussed in this subsection is very short. Here is what he writes:

> Since there are no arrows between variable [*sic.*] at different levels screened off by network variables, Gebharter claims that it is unclear over which causal paths probabilistic influence propagates between such higher- and lower-level variables [...]. True, there are no such arrows. But this is because, by assumption, screened-off variables influence each other, if at all, only via the network variables. When RCMC is satisfied, probabilistic influence propagates constitutionally (rather than causally) across the flattening's dashed arrows and causally across the same-level solid arrows. (Casini 2016, pp. 955f)

It seems to me that this reply does not solve the problem I have raised. It is clear, as Casini (2016) says in the quote above, that the fact that probabilistic influence between higher and lower level variables is only propagated via constitutional paths going through network variables holds due to an assumption made within the RBN approach, namely the recursive causal Markov condition. But saying that the problem I raised is a consequence of an assumption made does not solve the problem. The approach still does not tell us how a mechanism's input causally interacts with the mechanism's micro structure and how probability from the input to the output is propagated through this micro structure. As I have pointed out, observing that probability spreads without providing a causal story about the pathways from the input to the output variables does not suffice for mechanistic causal explanation. One cannot simply answer that the problem is no problem because it arises due to an assumption. In that case, the assumption made may not be the best choice and I would suggest to reconsider it.

6.3.2   Problem (ii): Impossibility of Bottom-Up Experiments

The second problem is even more striking than the first one discussed in Sect. 6.3.1. It is also connected to manipulation and control. One of the alleged merits of the mechanistic approach to explanation is that it is so close to scientific practice. Scientists, and especially biologists, carry out so-called top-down and bottom-up experiments to determine the mechanism of interest's constitutively relevant micro parts. The most prominent approach capturing this relation between manipulation and constitutive relevance in mechanisms is Craver's (2007a, b) mutual manipu-

---

Surgical (or ideal) interventions à la Pearl (2000) or Woodward (2003) would screen off T from W.

lability approach. In a nutshell, it states that a micro variable X is constitutively relevant for a macro variable Y if and only if X is a part of Y (in the mereological sense), X can be wiggled by manipulating Y (successful top-down experiment), and Y can be wiggled by manipulating X (successful bottom-up experiment). Though the details of Craver's mutual manipulability criterion are not uncontroversial (see Leuridan (2012) and Baumgartner and Gebharter (2016) for some serious problems with the mutual manipulability approach), it is commonly accepted that mechanisms typically allow for successful top-down and bottom-up experiments, and thus, that a mechanism's constitutively relevant parts can (more or less safely) be identified by carrying out such experiments. So far, so good.

Here comes the problem: The RBN approach entails the impossibility of bottom-up experiments by means of intervention variables. In other words: According to the RBN approach, it is not possible to represent an intervention on a micro variable as an intervention variable taking a certain value that influences the mechanism's output at the macro level. Let me demonstrate why this is so again by means of the water dispenser device. As shown in Sect. 6.2.1, the RBN formalism determines a unique probability distribution $\mathbf{P}$ over $\mathbf{V}$. In Sect. 6.2.2 it turned out that this probability distribution $\mathbf{P}$ for our example is $\mathbf{P}(T, B, D, S, H, W) = P^{n\downarrow}(T) \cdot P^{n\downarrow}(B) \cdot P^{n\downarrow}(D|T, B) \cdot P^{n\downarrow}(S|D) \cdot P^{n\downarrow}(H|D, S) \cdot P^{n\downarrow}(W|D)$. This probability distribution is, in turn, Markov-compatible with the DAG depicted in Fig. 6.7 (see Sect. 6.2.1), and so the two together form a BN. Let us now assume that there would be an intervention variable I_S for S (see Fig. 6.7 again). Then we can read off the BN directly that I_S is probabilistically independent of every one of the macro variables T, B, D, and W. Thus, there cannot be an intervention on S such that this intervention would have any influence on the RBN's top level. This does not only hold for S in case of the water dispenser, but for all micro variables of any mechanism represented by an RBN: The RBN approach leads to the consequence that bottom-up experiments by means of intervention variables are impossible, which stands in stark contrast to scientific practice as well as to core ideas which can be found in the mechanistic literature. Note that Casini et al. (2011, pp. 20f) explicitly intend the superiority/inferiority relation to capture constitutive relevance relations in mechanisms.

Fig. 6.7 The graph of a BN Markov-compatible with the water dispenser RBN's probability distribution $\mathbf{P}$ over $\mathbf{V} = \{T, B, D, S, H, W\}$. Any intervention variable I_S for S has to be probabilistically independent of all the variables at the RBN's macro level

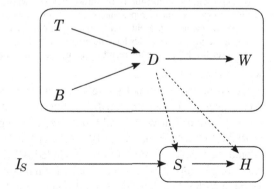

Casini (2016) agrees that the RBN approach cannot represent bottom-up experiments by means of intervention variables, but emphasizes that the effects of
interventions can still be computed. The post intervention distribution $\mathbf{P}^*(z)$ for
some variable or set of variables Z, when forcing another variable X to take a
certain value x by means of an intervention, can be computed with help of the
graph resulting from deleting all the arrows from X's causal parents pointing to
X (while leaving the arrows from X's direct superiors pointing to X intact) in the
RBN (cf. Casini 2016, sec. 3; Casini et al. 2011, pp. 12f). As a next step, one can
use the independencies implied by the resulting graph and the d-separation criterion
to compute the post intervention probability $\mathbf{P}^*(z)$. This method is intended to work
analogously to how post intervention distributions are computed according to Pearl
(1995, 2000, sec. 1.3.1). However, the method for computing post intervention
distributions in RBN models of mechanisms suggested by Casini et al. does not
lead to the intended results. It can be shown that their method regularly leads to
absurd consequences: Sometimes an intervention on a micro variable influences
not only this variable's effects, but also its causes (or even micro variables that
are not causally related at all to the intervention's target variable); sometimes
an intervention on a micro variable even influences the causes (or inputs) of the
corresponding network variable. I will explain this in detail in the next subsection.

Before I will highlight the third problem with the RBN approach, let me discuss
Casini's (2016) response to the second problem I have presented in this subsection.
Casini objects that my representation of an intervention on one of a mechanism's
micro variables by means of an intervention variable does not make much sense
within the RBN approach. The BNs with causal structure $S \longrightarrow H$ that are
D's values in our water dispenser example are intended to represent the possible
states of the mechanism represented by the network variable D. Interventions are
typically understood as external influences to the target system. But lower level
descriptions in RBN models of mechanisms are intended to only capture system
internal and constitutively relevant variables. Hence, so Casini argues, interventions
on a mechanism's micro parts cannot be represented as causes of micro variables:

> [...] RCMC is assumed to hold true of variables in [$\mathbf{V} = \{T, B, D, S, H, W, \}$], and not
> in the expanded set [$\mathbf{V}^+ = \{T, B, D, S, H, W, I_S\}$]. The reason is that RBNs are meant
> to represent decompositions of (properties of) wholes into (properties of) their parts;
> they are not meant to represent parts that do not belong to any whole [...]. The graph
> topology cannot represent such parts. Thus, one cannot read off the graph topology that
> such intervention variables have no effect. More generally, in an RBN, everything one gets
> at lower levels must be the result of (recursively) decomposing the top level. (Casini 2016,
> p. 957)

To be honest, I am not entirely sure whether Casini's (2016) claim is correct.
I can see, of course, that RBN models of mechanisms are meant to represent
decomposition. But why should this forbid adding external influences such as I_S?
If RCMC holds for the initial BN without the intervention variable I_S, then why
should it not hold anymore for the expanded BN (in Fig. 6.7)? Adding I_S as a direct
cause of S seems to be possible from a formal point of view: Adding an exogenous
variable X that is independent of all other variables Y not connected by a directed

path of continuous or dashed arrows to X should lead to a model that is Markov whenever the original model was.

However this may be, let us—just for the moment—assume that Casini's (2016) claim is correct and that there are reasons that make it impossible to represent interventions on a mechanism's micro parts by means of intervention variables in RBN models of mechanisms. The problem I have raised would still be unsolved. It is still not possible to represent bottom-up interventions by means of intervention variables. The reason would, however, be a different one. I have argued that if one would add intervention variables for micro parts, then there cannot be probability propagation from the intervention to the macro variables, which would be an absurd consequence. Hence, one cannot represent (successful) bottom-up experiments by means of intervention variables. If Casini's claim is right, on the other hand, and intervention variables for micro variables are not even possible in RBN models of mechanisms, then, of course, (successful) bottom-up experiments can also not be represented by means of intervention variables in RBN models of mechanisms.

6.3.3   Problem (iii): False Predictions Under Manipulation

To illustrate the third problem with Casini et al.'s (2011) RBN approach for representing mechanisms, I have to say a few words about how the effects of interventions are computed in ordinary causal Bayes nets when not representing interventions by means of intervention variables. Following Pearl (2000, sec. 1.3.1) I use the *do*-operator for formally distinguishing interventions from observations. $do(X = x)$ (or '$do(x)$' for short) means that X is forced to take value x by means of an intervention, while $X = x$ (or 'x' for short) just stands for observing that X takes value x. The main difference between intervention and observation is that probability propagation induced by intervention spreads only from the cause to the effect, while probability propagation induced by observation goes in both directions of the causal arrow. If X is a direct cause of Y and $X \longrightarrow Y$ is the only causal connection between X and Y, for example, intervening on X will typically lead to a probability change of Y (i.e., $P(y|do(x)) \neq P(y)$ will hold for some x, y), while intervening on Y will definitely not alter X's probability distribution (i.e., $P(x|do(y)) = P(x)$ will hold for arbitrarily chosen x, y). Observing that $X = x$, on the other hand, will typically lead to a change of Y's probability distribution (i.e., $P(y|x) \neq P(y)$ will hold for some x, y), while also observing $Y = y$ will typically lead to a change of X's probability distribution (i.e., $P(x|y) \neq P(x)$ will hold for some x, y).

Now computing post intervention distributions in ordinary causal Bayes nets works as follows: Let us assume that we want to set a variable X to a specific value x by means of an intervention and compute the effect of this intervention on Y. Y may be a variable or a set of variables different from X. As a first step, we delete all the causal arrows pointing at X from the graph of our causal Bayes net. One then uses the probabilistic independence information provided by the resulting

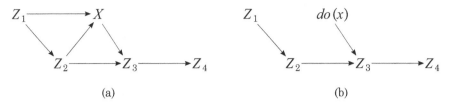

$$(a) \qquad\qquad\qquad\qquad\qquad (b)$$

Fig. 6.8 (**a**) Shows the original or unmanipulated graph, while (**b**) shows the graph after forcing X to value x by means of an intervention. The manipulated graph together with the d-separation criterion implies new independence constraints for the original observational distribution P

graph together with the causal Markov condition (or the equivalent d-separation criterion) to calculate the post intervention distribution $P(y|do(x))$. This corresponds to assuming that probability propagation induced by an intervention on X can only spread through the Bayes net along causal paths exiting X, i.e., along causal paths that do not feature an arrow pointing at X.

Here is an illustration how post intervention distributions can be computed: Assume our causal Bayes net of interest is the one whose graph is depicted in Fig. 6.8a. Assume further that we are interested in the probability of a certain Z_4-value z_4 when X takes a particular value x. The conditional probability $P(z_4|x)$, i.e., the probability of $Z_4 = z_4$ when observing that $X = x$, can be computed as follows:

$$P(z_4|x) = \sum_i P(z_4|x, z_{2_i}) \cdot P(z_{2_i}|x) \tag{6.3}$$

To compute the probability of $Z_4 = z_4$ when fixing X to value x by intervention, one first has to delete all arrows in Fig. 6.8a pointing at X. One then arrives at the graph in Fig. 6.8b. Now this graph together with the d-separation criterion tells us that $INDEP_P(X, Z_2)$, which implies $P(z_{2_i}|x) = P(z_{2_i})$ for all $z_{2_i} \in val(Z_2)$. Hence, we can compute the post intervention probability $P(z_4|do(x))$ on the basis of the original distribution together with the independence information provided by the modified graph in Fig. 6.8b and the d-separation criterion as follows:

$$P(z_4|do(x)) = \sum_i P(z_4|x, z_{2_i}) \cdot P(z_{2_i}) \tag{6.4}$$

Now Casini et al. (2011, pp. 12f.) suggest to adopt this method for computing post intervention distributions in RBNs as follows: As a first step, one has to delete all the causal parents of the variable X which one wants to manipulate, while all the arrows from X's direct superiors pointing to X should not be deleted. One then should, like in ordinary Bayes nets, use the independence relations induced by the resulting graph together with the recursive causal Markov condition to compute the post intervention distributions of interest. The motivation for deleting the arrows from variables in $Par(X)$ to X is the same as in ordinary BNs: Manipulating the cause will typically lead to a change in the effect, but manipulating the effect will not lead

to a change in the cause. However, the behaviors of the mechanisms represented by
an RBN's network variables are constituted by the behaviors of their corresponding
micro variables. Thus, manipulating a mechanism's micro variables should at least
sometimes lead to a change of the probability distribution of the corresponding
network variable. So we are not allowed to delete arrows from $DSup(X)$ to X when
intervening on X.

Let me briefly illustrate Casini et al.'s (2011) suggestion by means of an abstract
example: Assume we are interrested in a simple input output mechanism. I stands
for the mechanism's input, O for its output, and the network variable N represents
the mechanism mediating between certain inputs and outputs. So the RBN's top
level graph would be $I \longrightarrow N \longrightarrow O$. Now assume further that the mechanism's
causal micro structure is $A \longrightarrow B \longrightarrow C \longleftarrow D$. Then the RBN could be
represented by the graph in Fig. 6.9. Now let us assume that we want to compute
the post intervention probability of a certain O-value o when fixing B to a certain
value b by means of an intervention, i.e., $P(o|do(b))$. To compute $P(o|do(b))$, we
first have to delete all arrows exiting variables in $Par(X)$ and pointing at X. Hence,
we delete $A \longrightarrow B$, while we leave all other arrows untouched. The resulting graph
is depicted in Fig. 6.10.

Now we can use the independencies implied by the graph in Fig. 6.10 together
with the recursive causal Markov condition to compute $P(o|do(b))$ on the basis of
our original probability distribution $\mathbf{P}$. Since $P(o|b)$ equals $P(o, b)/P(b)$, we can
determine $P(o|do(b))$ by computing $P(o, b)$ and $P(b)$ on the basis of the modified
graph:

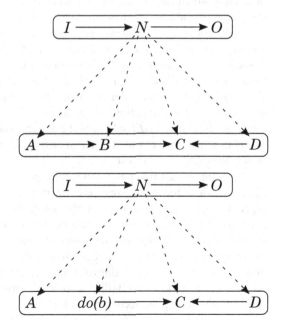

Fig. 6.9 RBN representation
of a simple input output
mechanism

Fig. 6.10 The graph one gets
by deleting all the causal
arrows pointing at the
manipulated variable B

$$\mathbf{P}(o, b) = \sum_i \mathbf{P}(n_i, o, b) = \sum_i \mathbf{P}(o|n_i) \cdot \mathbf{P}(b|n_i) \cdot \mathbf{P}(n_i) \qquad (6.5)$$

$$\mathbf{P}(b) = \sum_i \mathbf{P}(n_i, b) = \sum_i \mathbf{P}(b|n_i) \cdot \mathbf{P}(n_i) \qquad (6.6)$$

The strategy for computing post intervention distributions suggested by Casini et al. (2011) seems to work well for predicting a mechanism's output on the macro level when manipulating some of this mechanism's parts at the micro level. However, it will regularly lead to false predictions under interventions under the assumption of faithfulness. In a faithful Bayes net, every d-connection implies some (conditional) probabilistic dependence. So if our exemplary BN is faithful, then, since the structure resulting from deleting all and only the causal arrows exiting one of B's causal parents and pointing at B still features the dashed arrows $N \dashrightarrow A$, $N \dashrightarrow do(b)$, $N \dashrightarrow C$, and $N \dashrightarrow D$, several unexpected dependencies result. Path $I \longrightarrow N \dashrightarrow do(b)$ implies $DEP_\mathbf{P}(I, do(b))$ for some $b \in val(B)$, $A \dashleftarrow N \dashrightarrow do(b)$ implies $DEP_\mathbf{P}(A, do(b))$ for some $b \in val(B)$, and $do(b) \dashleftarrow N \dashrightarrow D$ implies $DEP_\mathbf{P}(D, do(b))$ for some $b \in val(B)$.

All three probabilistic dependencies under manipulation implied are highly alarming. That there is a B-value b such that forcing B to take this value b by means of an intervention leads to a change of I's probability distribution means that some manipulations of a mechanism's micro parts will lead to a change in the mechanism's input at the macro level. This would tantamount to accepting that intervening on the effect may change the cause. If, for example, our mechanism of interest would be a radio and I would stand for me fiddling with some of the radio's knobs and buttons, then your intervening on some of the radio's parts should definitely not have an influence on what I am doing with the radio's knobs and buttons. Also problematic is the dependence $DEP_\mathbf{P}(A, do(b))$. That there are some B-values b such that forcing B to take one of these values b by intervention has a probabilistic influence on A means that at least sometimes one can bring about the cause by bringing about the effect. This would totally blur the distinction between observation and manipulation, which is one of the main selling points of the whole causal net formalism. $DEP_\mathbf{P}(D, do(b))$ is maybe worse: Here some interventions on B lead to a change in D which is not an effect of B, and, moreover, not even a cause of B in the original graph's structure.

There are at least two possibilities to respond to my objection to Casini et al.'s (2011) method for computing post intervention distributions: First, one could argue that my objection hinges on the assumption that the modeled mechanisms are faithful (i.e., that every d-connection implies a dependence), but that many mechanisms—especially self regulatory mechanisms—will not satisfy this condition. Second, Casini et al. may suggest to modify their method in such a way that not only continuous arrows pointing at a variable X must be deleted when intervening on X, but also some of the dashed inter level arrows exiting X's direct superiors. I start with discussing the second possibility.

So could we not simply also delete all dashed arrows exiting N before computing post intervention probabilities? Unfortunately, this move would also be highly problematic. If we would delete all dashed arrows, I and B would be d-separated in the modified graph, and thus, $INDEP_\mathbf{P}(I, do(b))$ would hold for arbitrarily chosen $b \in val(B)$. That is what we want. But deleting $N \dashrightarrow B$ would, at the same time, d-separate B and O in the manipulated graph, which implies that manipulating B cannot lead to a change in O anymore. This can be generalized: Manipulating some of a mechanism's micro parts can under no circumstances lead to a difference for that mechanism's macro output, which contradicts scientific practice and our intuitions about causation and constitution. Hence, we do not want to delete $N \dashrightarrow B$. But if we do not delete $N \dashrightarrow B$, then the path $I \longrightarrow N \dashrightarrow do(b)$ in the resulting structure together with the faithfulness condition implies $DEP_\mathbf{P}(I, do(b))$ for some $b \in val(B)$, meaning that intervening on a mechanism's micro parts will at least sometimes have an effect on the mechanism's input at the macro level. If we want to avoid this absurd consequence and keep $N \dashrightarrow B$ intact, our only possibility is to delete $I \longrightarrow N$. So we want to keep $N \dashrightarrow B$ and to delete $I \longrightarrow N$.

Similar considerations apply to arrow $N \dashrightarrow C$. To see this, recall that N's values (which are BNs themselves) describe the possible states of the mechanism modeled. When we conditionalize on $do(b)$, we do not know in which state the mechanism is, i.e., we do not know which value N actually takes, meaning that we do not know which particular BN with structure $A \longrightarrow B \longrightarrow C \longleftarrow D$ we have to use to compute $do(b)$'s effect on C. Thus, we have to weight the causal influence of $do(b)$ on C in all BNs with structure $A \longrightarrow B \longrightarrow C \longleftarrow D$ by the probabilities of these BNs. This is the reason why we need the path $B \dashleftarrow N \dashrightarrow C$ intact.

Summarizing, the arrows $N \dashrightarrow B$ and $N \dashrightarrow C$ are required for generating correct predictions. So maybe we should only demand for deleting all those dashed arrows exiting N which do neither point at B, nor on an effect of B. In addition, as we saw above, we have to delete all arrows pointing at the network variable N. When doing so, we would end up with the graph depicted in Fig. 6.11. This move would block probability propagation over the problematic original paths $I \longrightarrow N \dashrightarrow B$, $A \dashleftarrow N \dashrightarrow B$, and $B \dashleftarrow N \dashrightarrow D$. Hence, the problematic dependencies $DEP_\mathbf{P}(I, do(b))$, $DEP_\mathbf{P}(A, do(b))$, and $DEP_\mathbf{P}(D, do(b))$ could be avoided.

So can Casini et al.'s (2011) method for computing post intervention probabilities finally be safed by the modification considered above? The answer to this question is

Fig. 6.11 The graph one gets by deleting all causal *arrows* pointing at the manipulated variable B, all causal *arrows* pointing at B's direct superior N, and all constitutional *arrows* which do neither point at B nor at an effect of B

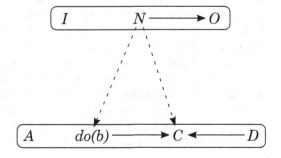

a negative one. This modified computation method directly leads to new problems. Here is the first problem: Conditionalizing on certain I-values should at least sometimes lead to probability changes of some N-values when fixing some micro variables by means of interventions. But this is excluded when we delete the arrow $I \longrightarrow N$; after deleting $I \longrightarrow N$, I is fully isolated from all the other variables. The second problem is that conditionalizing on A or on D while B is fixed to a certain value b by intervention should at least sometimes give us additional information on how probable certain states of the mechanism represented by N are. But also this is impossible according to the "improved" computation method, since after deleting $A \longrightarrow B$, $N \dashrightarrow A$, and $N \dashrightarrow D$ both A and D are d-separated from N by every subset of V. Summarizing, there seems to be no possible arrow deleting strategy that does not lead to false predictions.

Let us now have a closer look at the other possibility mentioned above Casini et al. (2011) have to respond to my initial objection. They could just insist that many mechanisms will not satisfy the faithfulness condition, which my argumentation requires to hold. My response would be that mechanisms, just like ordinary causal systems, can be expected to satisfy the faithfulness condition with very high probability if our measurement methods are precise enough. And even if only every third or fourth mechanism would be faithful, Casini et al.'s method would still regularly lead to false predictions. But I think I can say even more: Recall that violations of faithfulness require a very specific parameter fine-tuning (which is highly improbable). But the kind of non-faithful mechanisms for which Casini et al.'s prediction method may succeed requires very specific additional assumptions. Thus, mechanisms for which Casini et al.'s method succeeds can be expected to be even more rare than ordinary highly improbable non-faithful mechanisms.

Here are the details: As we saw above, mechanisms for which Casini et al.'s (2011) method would work have to feature paths $I \longrightarrow N \dashrightarrow B$, $A \dashleftarrow N \dashrightarrow B$, and $B \dashleftarrow N \dashrightarrow D$. These paths would have to produce very specific probabilistic properties: The conditional probabilities along the causal arrows must be chosen in such a way that every arrow is productive, but also in such a way that probability cannot be propagated from a path's starting point to its end point (or vice versa). Every arrow must be productive to avoid one of the problems discussed before. In particular, this assumption makes it possible that a mechanism's input still has an influence on the mechanism when some of this mechanism's micro variables are fixed by intervention. It is also required to make it possible that one achieves probabilistic information about the mechanism's state by learning about the actual behavior of some of the mechanism's micro parts while other micro parts are fixed by interventions. The assumption that the three paths mentioned do not propagate probabilistic influence between their starting points and their end points is required to avoid the second problem discussed earlier: It is required to avoid the consequence that manipulating a micro variable has an influence on other micro variables which are not effects of the micro variable one intervenes on. In addition, it is required to avoid the consequence that manipulating a mechanism's micro variables does lead to a change in the mechanism's input at the macro level.

Summarizing, there seems no promising way to go such that Casini et al.'s (2011) method for computing post intervention distributions does not regularly produce false predictions.

6.4 An Alternative Approach

6.4.1 The MLCM Approach for Modeling Mechanisms

In this subsection I develop an alternative approach—which I will call the *MLCM approach* (where 'MLCM' stands short for 'multi-level causal model')—for modeling mechanisms that does not share the problems discussed in Sect. 6.3 with Casini et al.'s (2011) RBN approach. I have first presented the basics of this approach in Gebharter (2014, sec. 4). Contrary to Casini et al. I do not represent mechanisms by means of a causal model's vertices, but rather by means of a causal model's arrows. I will use two types of causal arrows in my models: one-headed causal arrows ('$\longrightarrow$') and two-headed causal arrows ('$\longleftrightarrow$'). So I will not, as Casini et al., use BNs for representing the diverse levels of a mechanism, but causal models whose causal graphs do, contrary to causally interpreted Bayesian networks, not have to be DAGs. An arrow exiting X and pointing to Y ($X \longrightarrow Y$) in such a causal graph means that X is a direct cause of Y (relative to the graph's variable set V). A two-headed arrow between X and Y ($X \longleftrightarrow Y$) stands for the fact that there is a latent common cause of X and Y, i.e., a cause of X and Y not contained in the respective causal graph's variable set V. Note that causal models whose associated graphs contain one or more bi-directed causal arrows typically do not satisfy the causal Markov condition. However, causation in such causal graphs can in some way still be characterized by means of the causal Markov condition. But more on this later on.

First of all, note that my representation of mechanisms is much closer to mechanisms as they are described in the literature, *viz.* as input-output functions (Bechtel 2007; Craver 2007b, p. 45; Woodward 2013). The simplest model of a mechanism would contain just two variables X and Y and one causal arrow connecting them. Its causal graph would either be $X \longrightarrow Y$ or $X \longleftrightarrow Y$. In case we have a causal graph $X \longrightarrow Y$, X represents the mechanism's *input*, and Y its *output*. The causal arrow connecting X and Y stands for a (not further specified) mechanism at work. The causal graph $X \longrightarrow Y$ together with a probability distribution P over $V = \{X, Y\}$ then characterizes a mechanism, i.e., an input-output function. The causal graph tells us that X is a cause of Y, and the probability distribution P gives us the conditional probabilities $P(y|x)$ for all possible combinations of X- and Y-values x and y, respectively. If our causal graph is $X \longleftrightarrow Y$, on the other hand, then X and Y would both be outputs of the mechanism at work represented by this graph; this mechanism's input would in that case be unknown or not explicitly mentioned in the model.

What mechanists intend to provide are mechanistic explanations of certain input-output-behaviors by referring to the mechanism of interest's causal micro structure. To achieve a causal model useful for mechanistic explanation, we can assign a second causal model to the first one representing the mechanism's top level. This assigned causal model has to provide details about how, i.e., over which causal paths through the mechanism's micro structure, the mechanism's input produces its output. In case of the top-level model $X \longrightarrow Y$, such information could, for example, be provided by assigning a model with the associated graph $X \longrightarrow Z \longrightarrow Y$, where Z represents a micro part of the mechanism at work. However, it must be guaranteed that the top-level model and the assigned model providing information about the mechanism's micro structure fit together with respect to the causal as well as to the probabilistic information they store. This can be done by the following notion of a *restriction* (cf. Gebharter 2014, p. 147). This notion of a restriction is a modified version of Steel's (2005, p. 11) notion of a restricted graph supplemented by specific conditions for structures containing bi-directed arrows[6]:

Definition 6.2 (restriction) $\langle V, E, P \rangle$ is a restriction of $\langle V', E', P' \rangle$ if and only if

(a) $V \subset V'$, and
(b) $P' \uparrow V = P$, and
(c) for all $X, Y \in V$:

(c.1) If there is a directed path from X to Y in $\langle V', E' \rangle$ and no vertex on this path different from X and Y is in V, then $X \longrightarrow Y$ in $\langle V, E \rangle$, and

(c.2) if X and Y are connected by a common cause path π in $\langle V', E' \rangle$ or by a path π free of colliders containing a bidirected edge in $\langle V', E' \rangle$, and no vertex on this path π different from X and Y is in V, then $X \longleftrightarrow Y$ in $\langle V, E \rangle$, and

(d) no causal path not implied by (c) is in $\langle V, E \rangle$.

The intuition behind a restriction as defined above is to allow for marginalizing out variables of a causal model richer in details in such a way that no probabilistic or causal information provided by this model gets lost. Condition (a) guarantees that the restricted model's set of variables V contains fewer variables than the original model's set of vertices V', condition (b) that all the probabilistic information captured by the original model is provided when restricting it to V, and (c) that also causal information is conserved. In particular, (c.1) tells us when we have to draw a one-headed arrow in the restricted model, while (c.2) tells us when we have to draw a two-headed arrow: Draw a one-headed arrow coming from X and pointing to Y in the restricted model's graph if X is a cause of Y in the original model, but all intermediate causes lying on this path are not represented in the restricted model's

---

[6]As Clark Glymour (personal communication) pointed out to me, the marginalization method provided by Definition 6.2 is actually some kind of a "slim" version of Richardson and Spirtes' (2002) mixed ancestral graph representation which was developed for latent variable models in the possible presence of selection bias (see also Spirtes et al. 1999).

$$X \longleftrightarrow Y \longleftarrow Z \longleftarrow U \longrightarrow W$$

Fig. 6.12 Causal graph of an exemplary causal model for illustration how marginalizing out variables by appliance of Definition 6.2 works

variable set anymore. Draw a bi-directed arrow between X and Y in the restricted model's graph if at least one of the following scenarios is the case: (i) X and Y are connected by a common cause path in $\langle V', E' \rangle$ and all variables lying on this path are not contained in V; (ii) X and Y are connected by a path that does contain a bi-directed arrow, no subpath of the form $Z_k @\longrightarrow Z_l \longleftarrow @ Z_m$,[7] and no variable lying on this path is contained in the restricted model's vertex set V. Condition (d) tells us that the causal paths we have to draw because of (c.1) and (c.2) are all the causal paths in the restricted model.

Let me briefly illustrate how Definition 6.2 can be used to marginalize out variables on a concrete exemplary causal model (whose causal graph is depicted in Fig. 6.12). If one would decide to marginalize out U, then the resulting causal model's graph would, according to Definition 6.2, be $X \longleftrightarrow Y \longleftarrow Z \longleftrightarrow W$. If the original model would be restricted to variable set $\{X, Y, U, W\}$, the resulting topology would be $X \longleftrightarrow Y \longleftarrow U \longrightarrow W$. When marginalizing out Z and U, we would get the graph $X \longleftrightarrow Y \longleftrightarrow W$, and when marginalizing out Y we would get $X \quad Z \longleftarrow U \longrightarrow W$ (without any causal connection between X and Z) as the restricted model's causal graph.

Another nice property of the notion of a restriction in Definition 6.2 is that it preserves *d*-connection information when going from the original model to any one of its restrictions:

Theorem 6.1 *If $\langle V, E, P \rangle$ is a restriction of $\langle V', E', P' \rangle$, then for all $X, Y \in V$ (with $X \neq Y$) and $M \subseteq V \backslash \{X, Y\}$: If X and Y are d-connected given $M \subseteq V \backslash \{X, Y\}$ in $\langle V', E' \rangle$, then X and Y are d-connected given M in $\langle V, E \rangle$.*

Proof Assume $\langle V, E, P \rangle$ is a restriction of $\langle V', E', P' \rangle$. Also assume that X and Y are two arbitrarily chosen variables in V (with $X \neq Y$) and that M is a subset of $V \backslash \{X, Y\}$. Now suppose that X and Y are *d*-connected given $M \subseteq V \backslash \{X, Y\}$ in $\langle V', E' \rangle$, meaning that there is a path $\pi' : X @\!-\!@ \ldots @\!-\!@ Y$ in $\langle V', E' \rangle$ via which X and Y are *d*-connected given M in $\langle V', E' \rangle$. Now there are two possible cases: Either (i) π' does not feature colliders, or (ii) π' features colliders. If (i) is the case, then π' has one of the following forms:

(i.i) $X \longrightarrow\!\!\longrightarrow Y$
(i.ii) $X \longleftarrow\!\!\longleftarrow Y$
(i.iii) $X \longleftarrow\!\!\longrightarrow Y$
(i.iv) $X \longleftarrow\!\!\longleftarrow Z_i \longleftrightarrow Z_j \longrightarrow\!\!\longrightarrow Y$ (where Z_i may be identical to X and Z_j may be identical to Y)

---

[7] '$@\!\longrightarrow$' is a meta symbol for '$\longrightarrow$ or $\longleftrightarrow$'.

Note that in all of these cases M cannot contain any variable lying on π'. Now there are, again, two possible cases: Either (A) no variable $Z \neq X, Y$ on π' is in $V' \backslash V$, or (B) some variables $Z \neq X, Y$ on π' are in $V' \backslash V$. If (A) is the case, then, according to condition (c) of Definition 6.2, there will be a path π in $\langle V, E \rangle$ that is identical to π'. Hence, X and Y will be d-connected given M in $\langle V, E \rangle$.

Assume (B) is the case. Then condition (c) of Definition 6.2 applied to π' implies: If (i.i), then there will be a directed path π from X to Y in $\langle V, E \rangle$ not going through M. If (i.ii), then there will be a directed path π from Y to X in $\langle V, E \rangle$ not going through M. If (i.iii), then there will be a common cause path π or a path free of colliders containing a bidirected edge between X and Y in $\langle V, E \rangle$ not going through M. If (i.iv), then there will be a path π free of colliders containing a bidirected edge between X and Y in $\langle V, E \rangle$ not going through M. In all of these cases, X and Y will be d-connected given M in $\langle V, E \rangle$. Note that in case (A) as well as in case (B) the path π features an arrow head pointing at X if and only if π' features an arrow head pointing at X and that the path π features an arrow head pointing at Y if and only if π' features an arrow head pointing at Y.

Now let us assume that (ii) is the case. Assume Z is the only collider on π'. Since π' d-connects X and Y given M, (A) Z is in M, or (B) there is an effect Z' of Z that is in M. Assume (A) is the case. There must be two subpaths π'_X and π'_Y of π' such that Z is d-connected with X given M via π'_X, Z is d-connected with Y given M via π'_Y, and π' is the concatenation of π'_X and π'_Y. Since π'_X and π'_Y do not feature colliders, they are either directed paths, common cause paths, or paths featuring one bidirected edge. As we saw from the investigation of case (i) above, it follows that there are corresponding paths π_X and π_Y in $\langle V, E \rangle$. π_X features an arrow head pointing at X if and only if π'_X features an arrow head pointing at X, π_X features an arrow head pointing at Z if and only if π'_X features an arrow head pointing at Z, π_Y features an arrow head pointing at Z if and only if π'_Y features an arrow head pointing at Z, and π_Y features an arrow head pointing at Y if and only if π'_Y features an arrow head pointing at Y. Thus, X and Y will be d-connected given M via π in $\langle V, E \rangle$. In case π' features more than one collider, then the same reasoning as above can be applied. One just has to cut π' in more than two subpaths. These paths are the longest subpaths of π' not featuring a collider themselves.

Now assume that (B) is the case. Then we have two collider free paths π'_X and π'_Y such that X and Z' are d-connected given M via π'_X, while Y and Z' are d-connected given M via π'_Y. (Note that π'_X and π'_Y will not be subpaths of π' this time, because they both feature $Z \longrightarrow \longrightarrow Z'$, which is not a subpath of π'.) By a similar reasoning as in case (A) above it follows that there must be corresponding paths π_X and π_Y in $\langle V, E \rangle$: π_X features an arrow head pointing at X if and only if π'_X features an arrow head pointing at X, π_X features an arrow head pointing at Z if and only if π'_X features an arrow head pointing at Z, π_Y features an arrow head pointing at Z if and only if π'_Y features an arrow head pointing at Z, and π_Y features an arrow head pointing at Y if and only if π'_Y features an arrow head pointing at Y. Let π be the concatenation of π_X and π_Y. Then X and Y will be d-connected given M via π in $\langle V, E \rangle$.

If π' features more than one collider, then the same reasoning as above can be applied. One just has to identify more than two subpaths of π'. Here is one possibility to do it: Start with X. Check whether the next collider Z_1 on π' is in M. If $Z_1 \in M$, then the first path π'_1 is the part of π' between X and Z_1. If $Z_1 \notin M$, then the first path p_1 is the concatenation of the part of π' between X and Z_1 and the directed path from Z_1 to Z_1's effect $Z'_1 \in M$. Then we apply the same procedure starting with Z_1 in case $Z_1 \in M$ and starting with Z'_1 in case $Z_1 \notin M$. Repeat this procedure till π's endpoint Y is reached. □

A nice and direct consequence of Theorem 6.1 is that every restriction of a causal model satisfying the d-connection condition will also satisfy the d-connection condition.

With Definition 6.2 we have now a tool at hand by whose means we can connect causal models representing different levels of one and the same mechanism. With its help we can define the following notion of a *multi-level causal model* (MLCM; cf. Gebharter 2014, p. 148) which, as I will try to demonstrate later on, can nicely capture the hierarchic structure of nested mechanisms without falling victim to the problems discussed in Sect. 6.3 Casini et al.'s (2011) RBN approach has to face:

Definition 6.3 (multi-level causal model) $\langle M_1 = \langle V_1, E_1, P_1 \rangle, \dots, M_n = \langle V_n, E_n, P_n \rangle \rangle$ is a multi-level causal model if and only if

(a) $M_1, \dots, M_n$ are causal models, and
(b) every M_i with $1 < i \le n$ is a restriction of M_1, and
(c) M_1 satisfies CMC.

According to condition (a), M_1 to M_n are causal models representing causal systems which may include one or more mechanisms. Such mechanisms are represented by one- or two-headed causal arrows together with the variables standing at these arrows' heads and tails (as described before) in the respective models. (b) uses the notion of a restriction introduced in Definition 6.2 to guarantee that the diverse models' associated probability distributions as well as their causal graphs fit together in the sense explained above. Condition (c), finally, assures that causal relationships in the MLCM's models are characterized by the causal Markov condition. So what we assume here is that for every causal model describing a mechanism at a higher level, there has to be an underlying causal structure satisfying CMC such that marginalizing out variables of this causal structure results in the causal model representing the mechanism at the higher level. This is the mechanistic version of a basic assumption made within the causal nets approach: Every probability distribution over a set of variables can be explained by some (maybe larger) underlying causal structure satisfying the causal Markov condition (cf. Spirtes et al. 2000, pp. 124f).

Though an MLCM $\langle M_1 = \langle V_1, E_1, P_1 \rangle, \dots, M_n = \langle V_n, E_n, P_n \rangle \rangle$ can be used to model nested mechanisms, it does not directly tell us the hierarchical order of the represented mechanisms different from M_1. Recall that Definition 6.3 just tells us that every M_i appearing later in the ordering is a restriction of M_1. However, there is an easy way to get information about the ordering of the causal models M_i in

$\langle M_1 = \langle V_1, E_1, P_1 \rangle, \ldots, M_n = \langle V_n, E_n, P_n \rangle \rangle$. This can be done by constructing an MLCM's *level graph G*. This level graph is some kind of meta-graph providing information about the hierarchical order of the models of the MLCM of interest. It is defined as follows (cf. Gebharter 2014, p. 149):

Definition 6.4 (level graph) $G = \langle V, E \rangle$ is an MLCM $\langle M_1 = \langle V_1, E_1, P_1 \rangle, \ldots, M_n = \langle V_n, E_n, P_n \rangle \rangle$'s level graph if and only if

(a) $V = \{M_1, \ldots, M_n\}$, and
(b) for all $M_i = \langle V_i, E_i, P_i \rangle$ and $M_j = \langle V_j, E_j, P_j \rangle$ in V: $M_i \longrightarrow M_j$ in G if and only if $V_i \subset V_j$ and there is no $M_k = \langle V_k, E_k, P_k \rangle$ in V such that $V_i \subset V_k \subset V_j$ holds.

Condition (a) guarantees that all causal models of the MLCM are vertices of this MLCM's level graph. Condition (b) provides information about the level graph's topology: It instructs us to draw an arrow from a causal model M_i to a causal model M_j ($M_i \longrightarrow M_j$) if and only if two conditions are satisfied. First, all variables in M_i's variable set V_i have to be also elements of M_j's vertex set V_j. Second, there must not be any causal model M_k in V such that M_k's variable set V_k is a proper superset of V_i and a proper subset of V_j. What we ultimately get by appliance of Definition 6.4 to an MLCM is, thus, a DAG G whose directed paths correspond to the proper subset relation in set theory.

When using MLCMs to model mechanisms, an MLCM's level graph explicitly provides all the information about the hierarchic ordering of causal models included in the MLCM which there is. It gives us a strict order among the models of the MLCM. If the MLCM's level graph features a directed path from causal model M_i to causal model M_j, then this indicates a difference in levels: M_j would be at a higher level than M_i in case $M_i \longrightarrow \ldots \longrightarrow M_j$ in G. If, in addition, we observe that a causal model M_k different from M_i and M_j lies on this path $M_i \longrightarrow \ldots \longrightarrow M_j$, then M_k represents a mechanism at a level between those represented by M_i and M_j. Note that, according to Definition 6.3, an MLCM does always have exactly one lowest level causal model, *viz.* M_1, and that every causal model different from this model M_1 is a restriction of M_1. Thus, an MLCM's level graph does always contain exactly one vertex, *viz.* M_1, such that for every vertex different from M_1 there is a directed path from this vertex to X_1. (See Fig. 6.13 for a graphical illustration.)

For further illustration how the hierarchical order of nested mechanisms can be read off a level graph, let us take a brief look at the exemplary MLCM $\langle M_1, \ldots, M_5 \rangle$'s level graph depicted in Fig. 6.13. (Note that $M_2, \ldots, M_5$ are restrictions of M_1.) We can choose any two causal models M_i and M_j of $\{M_1, \ldots, M_5\}$ and ask the level graph to tell us their hierarchical relationship. Since $M_5 \longrightarrow M_4 \longrightarrow M_2 \longrightarrow M_1$ is a directed path in the level graph, M_5 represents a higher level than M_4, which represents a higher level than M_2, which, in turn, represents a higher level than M_1. Because the level graph features the directed path $M_5 \longrightarrow M_3 \longrightarrow M_1$, we know that M_5 stands for a higher level than M_3, which, again, represents a higher level than M_1. If we take a look at M_2 and M_3, we find that these two vertices are not connected by a directed path in the level graph, meaning that the MLCM does not provide information about whether M_2 is located at a higher or at a lower level

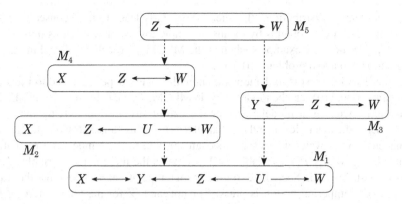

Fig. 6.13 Level graph of an exemplary MLCM. The level graph's edges are represented by *dashed arrows*. The *continous arrows* represent direct causal connections within a causal model M_i (with $1 \le i \le 5$)

than M_3. It would even be possible that both models represent structures at the same level. The same holds for M_3 and M_4. However, the MLCM's level graph tells us that all three models, M_2, M_3, and M_4 are located somewhere between M_1 and M_5 in the hierarchy.

6.4.2 MLCMs vs. RBNs

In this subsection I compare the MLCM approach for modeling mechanisms developed in Sect. 6.4.1 with Casini et al.'s (2011) RBN approach discussed in Sect. 6.2. First of all, note that the MLCM approach shares the main merit of the RBN approach: It provides a probability distribution over all the variables representing the modeled mechanisms' behaviors and the possible behaviors of their respective parts at all levels represented in the model. In case of an RBN, this probability distribution is **P**, in case of an MLCM, it is the probability distribution P_1 of the lowest level causal model M_1 of which all other models of the MLCM are restrictions. Both probability distributions, **P** and P_1, can be used for prediction across the levels of the mechanism(s) represented.

However, recall the three problems the RBN approach has to face which I have presented in Sect. 6.3: Problem (i) was that RBNs do not graphically represent the causal connection of a mechanism's input and output variables to this mechanism's causal micro structure. This is problematic when it comes to mechanistic explanation as well as to questions concerning manipulation and control. Problem (ii) was that the RBN approach leads to the consequence that the representation of bottom-up experiments by means of intervention variables is impossible. Problem (iii) was that the RBN approach also cannot represent interventions adequately as arrow breaking interventions. Though Casini et al. (2011) suggest a method for computing post intervention distributions, this method regularly leads to false consequences. I illustrated all three problems by means of simple toy examples. Let us now assess by means of the same examples whether the MLCM approach developed in the last subsection can avoid problems (i)–(iii).

What I will do first is to demonstrate how the water dispenser (used to illustrate problems (i) and (ii)) can be modeled by an MLCM. This can be done by an MLCM containing two causal models: M_2 for the causal structure underlying the water dispenser at the macro level, and M_1 that tells us how this causal structure is causally connected to the water dispenser's mechanism, i.e., to the temperature regulation unit. $M_2 = \langle V_2, E_2, P_2 \rangle$'s variable set V_2 contains the three macro variables T, B, and W. T stands, as in Sect. 6.2.2, for the room temperature, B for whether the water dispenser's tempering button is pressed or not, and W for the temperature of the water dispensed. M_2's causal structure is $T \longrightarrow W \longleftarrow B$. Note that no variable D is used to represent the mechanism at the macro level as in Casini et al.'s (2011) RBN approach. The mechanism is, instead, represented by the subgraph $T \longrightarrow W \longleftarrow B$. T and B represent the mechanism's input, W stands for its output, and the causal arrows for the mechanism at work. The probability distribution P_2 tells us how certain inputs determine the probabilities of certain outputs. $M_1 = \langle V_1, E_1, P_1 \rangle$'s variable set V_1 is $\{T, B, S, H, W\}$, its causal structure consists of the two directed causal paths $T \longrightarrow S \longrightarrow H \longrightarrow W$ and $B \longrightarrow H$. $S \longrightarrow H$ describes the temperature regulation unit's causal micro structure, and the rest of the model's graph shows how the two micro variables S and H are causally connected to the mechanism's input (T and B) and output (W). M_2 is assumed to be a restriction of M_1, and thus, condition (b) of Definition 6.2 rules that $P_1 \uparrow V_2 = P_2$. The MLCM's level graph is depicted in Fig. 6.14.

We can observe now that the MLCM representation definitely allows for avoiding problem (i): Its lower level model M_1 gives detailed information about how the mechanism's input variables (T and B) are causally connected to the mechanism's micro variables (S and H), and how these micro variables cause the output of the mechanism (W). So the MLCM can be used for mechanistic explanation. Let us see,

Fig. 6.14 Level graph of an MLCM representing the water dispenser

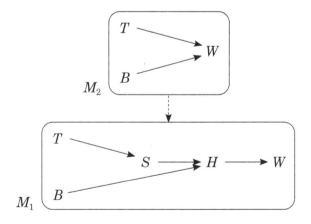

for example, how it can explain the phenomenon discussed in Sect. 6.3.1 the RBN model could not explain, *viz.* that T and W are independent when the tempering button B is not pressed ($B = 0$). The MLCM explains this phenomenon by providing just the thing required for explanation, *viz.* the causal structure $\langle V_1, E_1 \rangle$. It is also this causal structure that provides answers to certain questions about manipulation and control: It tells us, for example, that T's influence on W can be altered by manipulating the intermediate causes between T and W, i.e., S and/or H, by means of stochastic interventions.

The MLCM does also allow for a representation of bottom-up experiments by means of intervention variables, and thus, does not share problem (ii) with Casini et al.'s (2011) RBN approach. We can just add an intervention variable for any one of the micro variables, e.g., an intervention variable I_S for S like in standard causal models as they are, for example, discussed in Spirtes et al. (2000). Such an intervention on S can, of course, have an effect on the probability distribution over the mechanism's output, i.e., the macro variable W.

Before I show how MLCMs can deal with problem (iii), I want to briefly discuss a comment of Casini (2016). In a footnote at the end of section 1 of his paper Casini complains about how I compare the MLCM approach and the RBN approach by means of the water dispenser example. He writes:

> Gebharter contrasts the virtues of this MLCM with an RBN of the 'same' mechanism (2014, 142-3). However, this is somewhat misleading. Gebharter's RBN is defined over a larger variable set, which includes a network binary variable D, superior to S and H, caused by T and B, and causing W. It is obvious that his RBN cannot represent the same mechanism as his MLCM. On the assumption that the RBN is faithful, it should be possible to order the RBN's flattening (Gebharter 2014, 144), call it M_0, as *prior* with respect to M_1—since M_1's variable set V_1 is $\{V_0 \backslash D\}$. However, M_1 is incompatible with the restriction of M_0 obtained by marginalizing out D, call this M_{1*}. (M_{1*} would contain $S \longleftrightarrow H, S \longleftrightarrow W, H \longleftrightarrow W$ and $B \longrightarrow S$. Instead, M_1 contains $S \longrightarrow H, H \longrightarrow W$ and $B \longrightarrow H$.) Thus, rather than one model being a correct representation and the other being a wrong representation of one and the same mechanism, the two models represent different mechanisms, and are thus are [*sic.*] not directly comparable. (Casini 2016, sec. 1)

It seems to me that several things went wrong here. First of all, there is definitely not more than one mechanism. The mechanism is the thing out there in the world, the models (built within the frameworks of the two different modeling approaches) are our reconstructions of this mechanism. Since the MLCM approach represents mechanisms by means of arrows and the RBN approach represents mechanisms by means of network variables, the two models have to be different. But this does not imply that there are two mechanisms. There are just two different models of one and the same mechanism. And, of course, not every model is an adequate representation of a mechanism. It seems to me that the MLCM I have suggested is the most adequate representation within the MLCM approach, while the RBN I have suggested is the most adequate representation within the RBN approach. The mechanism that mediates between the input T and B and the output W is represented by the arrows in M_2 within the MLCM approach. M_1 tells one more about this mechanism's micro structure and how it is connected to its input and output. Within the RBN approach, on the other hand, I have to represent the mechanism by means of a network variable D. The mechanism's micro structure must be given by D's values (i.e., the BNs with structure $S \longrightarrow H$).

As far as I can see, it does not make much sense to order Casini's (2016, sec. 1) M_0 as "prior with respect to M_1—since M_1's variable set is $\{V_0 \backslash D\}$" in an MLCM. It does not even make sense to have a network variable D in any model of an MLCM. That one can marginalize out D from the structure M_0 by means of the method I have suggested does not mean that M_0 or the resulting model M_{1*} is an adequate representation of a mechanism within the MLCM approach. Why should M_{1*} be an adequate representation if already M_0 is not adequate? And, of course, M_{1*} is incompatible with M_1, just as Casini says. But this cannot be read as evidence that the two models cannot be compared. It can simply be explained by the fact that M_1 is an adequate representation of the mechanism and its connection to the mechanism's input and output variables, while neither M_0 nor M_{1*} is.

Last but not least, let us see how the MLCM approach can avoid problem (iii). The abstract example used to illustrate this problem had $I \longrightarrow N \longrightarrow O$ as its causal macro structure and $A \longrightarrow B \longrightarrow C \longleftarrow D$ as its micro structure. I stood for the mechanism's input, O for its output, and the network variable N represented the mechanism at work. Within the MLCM approach we do not need a network variable N. Instead we represent the mechanism at the top level by the simple input-output structure $I \longrightarrow O$. At the micro level we have to connect structure $A \longrightarrow B \longrightarrow C \longleftarrow D$ somehow to I and O. Let us assume, for example, that I directly causes A and D, while C directly causes O. Then we get the MLCM whose level graph is depicted in Fig. 6.15.

Now note that this MLCM can be used to correctly compute post intervention distributions. One of the problems of the RBN representation was that intervening on B sometimes influences I's probability distribution. When we intervene on B in our lower-level model M_1, we have to delete all incoming arrows, i.e., we have to delete $A \longrightarrow B$ and get the graph depicted in Fig. 6.16. Now the problem for

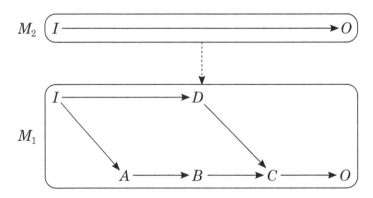

Fig. 6.15 Level graph of an MLCM representing the abstract example used for illustrating problem (iii)

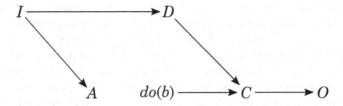

Fig. 6.16 Manipulated graph of model M_1

the RBN approach was that B was still d-connected to I in the modified graph. In the modified graph of our model M_1, on the other hand, B and I are clearly d-separated. Hence, $INDEP_{P_1}(I, do(b))$ has to hold for arbitrarily chosen $b \in val(B)$ and manipulating a mechanism's micro variables does not lead to a change of the mechanism's input. For RBNs an analogous problem arose w.r.t. A and D. Observe that B is also d-separated from A and D in our modified graph. Thus, the problematic dependencies $DEP_{P_1}(A, do(b))$ and $DEP_{P_1}(D, do(b))$ cannot arise, meaning that manipulating a mechanism's micro parts does not lead to a change to some of these micro parts' causes or even to micro variables that are not even causally connected to B.

6.4.3   Recent Objections Against the MLCM Approach

Casini (2016) has meanwhile responded to my criticism of the RBN approach for modeling mechanisms brought forward in Gebharter (2014). Casini also formulates several objections against my MLCM approach in that paper. In this subsection, I defend the MLCM approach against these objections.

Here is Casini's (2016) first worry. The MLCM he is speaking about in the following quote is, again, the water dispenser MLCM.

> First, it is not clear if MLCMs adequately represent mechanistic decompositions. High-level [*sic.*] causal models in a MLCM [...] are just more coarse-grain [*sic.*] representations of one and the same mechanism [...] such that some of the information in [the lower-level model] is *missing* at the higher level, as the term 'restriction' suggests. Is, for instance, $T \longrightarrow S \longrightarrow H \longrightarrow W$ a mechanistic decomposition of $T \longrightarrow W$, although *entities* and *properties* involved are the same at both levels, and only some *activities* (or relations) are different? Perhaps this counts as a different, equally legitimate, notion of decomposition, call it *decomposition**. The question is: How intuitive is *decomposition**? (Casini 2016, sec. 2)

First of all, I never claimed that I have provided an account of mechanistic decomposition. Such an account would have to come with a criterion that tells us how to decompose mechanisms. I have not given such a criterion. One may think that the notion of a restriction I introduced is intended as such a criterion. However, it clearly is not. Assume we have a model M_1 that correctly describes the structure of a mechanism and how this mechanism is connected to some input and output variables. Then clearly not every possible restriction M_2 is an adequate description of this mechanism at the macro level. And vice versa: Assume we have a macro level model M_2 adequately representing a mechanism. Then it is not guaranteed that every model M_1 we can find such that M_2 is a restriction of M_1 adequately describes the mechanism's micro structure and how it is connected to its in- and outputs. Let me emphasize again that the MLCM approach is a modeling approach, and not an approach of mechanistic decomposition. This means that it can be used to model mechanisms given one already knows all the relevant details, i.e., which parts of the systems are the components of the mechanism and how they are causally connected to each other etc. It is the same with the RBN approach. RBNs can be used to model mechanisms given that one already knows the relevant parts and the causal and constitutive relationships.

Casini (2016) asks whether $T \longrightarrow S \longrightarrow H \longrightarrow W$ is a mechanistic decomposition of $T \longrightarrow W$. It clearly is not. $T \longrightarrow W$ does not represent a mechanism, it just represents a part of a mechanism. The mechanism would be fully described at the macro level by *both* arrows in M_2's structure $T \longrightarrow W \longleftarrow B$. Now it makes sense to speak of mechanistic decomposition, if one wants to. However, the formalism alone (i.e., the formalism without any formalism-external additional information) does not tell us anything about decomposition. *If* the MLCM $\langle M_1, M_2 \rangle$ correctly represents the water dispenser mechanism and *if* there are no components of the mechanism different from S and H, *then*, I guess, one could say that the mechanism represented by the arrows in $T \longrightarrow W \longleftarrow B$ has correctly been decomposed into $S \longrightarrow H$. But let me say it again: The formalism does not provide a criterion for decomposition. We have to answer the decomposition question independently of the MLCM approach.

But the main worry Casini (2016) seems to have is that all of the variables of the higher level model also appear in the lower level model in MLCMs. So how can the lower-level model then be a decomposition of the higher level model? The simple answer is that it cannot be a decomposition. In the water dispenser MLCM, for example, not all variables in M_1 are parts of the mechanism represented by the arrows of M_2. Only S and H are. Now Casini could complain that the MLCM approach does not tell us that S and H are the relevant parts of the mechanism. To illustrate this, one could simply add arbitrarily chosen new variables to M_1; for example, one could add a variable A standing for an air conditioner which regulates the temperature of the room the water dispenser stands in. A would be exogenous and a direct cause only of T. The notion of a restriction does not exclude this. Here is my response: Of course, the formalism does not exclude adding A to M_1. But clearly not every variable we add to M_1 in such a way that M_2 is still a restriction of M_1 is a relevant part of the mechanism of interest. We have to know before constructing the model which variables are relevant parts and which ones are not. And we have to know in advance which subgraphs of M_1 represent mechanisms represented by arrows in M_2.

Casini's (2016) second worry is about explanation. He writes:

> Second, it is not clear if MLCMs adequately represent mechanistic explanations. One may concede that there is a legitimate sense in which one explains the relation between, say, the room temperature T and the water temperature W by blowing up the process from the former to the latter and uncovering the mediating role of the sensor S and the heater H. However, this sort of explanation is different from the equally legitimate explanation whereby one redescribes the cancer mechanism C [...] into more fine-grain terms, and uncovers the role of damage G and response D. G and D have an obvious mechanistic role. Instead, S and H seem to have an *etiological* role. Perhaps S and H still explain mechanistically, according to some different notion of mechanistic explanation, call it *explanation**. But just how intuitive is *explanation**? (Casini 2016, sec. 2)

The cancer mechanism Casini (2016) speaks of in the quoted paragraph is modeled by an RBN whose top level graph is $C \longrightarrow S$ and whose lower level graph is $G \longrightarrow D$. C is a network variable whose direct inferiors are G and D. C stands for whether some tissue in an organism is cancerous, S for the survival after 5 years, G for unregulated cell growth resulting from mutations, and D for cell division. This RBN can be used to mechanistically explain two different things. One could use G and D to mechanistically explain the output S. In that case, the relevant parts G and D would have an etiological role, just as S and H have for the explanation of the water dispenser's output W. The second possibility would be to use G and D to mechanistically explain C in the cancer example. In that case, the mechanism's parts would be used to explain the behavior of the mechanism as a whole and not the behavior of its output. In this explanation G and D would have a diachronical role. Since there is no variable representing the mechanism at the macro level in the water dispenser MLCM, S and H cannot have a diachronical role when one uses the MLCM for explanation. This is the difference Casini seems to point to: Micro variables in MLCMs can only be used for etiological mechanistic explanation, while micro variables in RBNs can be used for etiological as well as for diachronical explanation.

I do agree that one might sometimes be more interested in explaining the behavior of a system in terms of its parts which are relevant for that behavior rather than in explaining a certain output of that system. And I also have to accept, as Casini (2016) seems to think, that the water dispenser MLCM (in its current form) cannot be used for a diachronic explanation of the behavior of the mechanism mediating between the input variables T and B and the output variable W. The MLCM's top level model M_2 does not have a variable which would allow the representation of this mechanism. I think, however, that one cannot infer from this observation that MLCMs in general cannot be used for diachronical explanation. If one is more interested in a higher level description of the mechanism mediating between the water dispenser's in- and output, one could, for example, simply replace the output variable W at the macro level by a variable describing this mechanism (at the macro level). Let us call this variable M. M would then play the same role in the MLCM as the output variable O before. Our new MLCM would have the structure depicted in Fig. 6.17a. The mechanism would then have to be understood as an input output function, where the output is the behavior of the system as a whole.

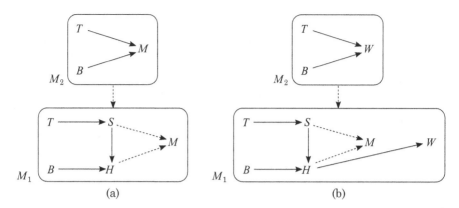

Fig. 6.17 (**a**) MLCM that allows S and H to play a diachronical role when used for explanation of M. M plays the role of an output variable. (**b**) MLCM that allows S and H to play a diachronical role when used for explanation of M and an etiological role when used for explanation of W. The *arrows* in M_2 still represent the mechanism mediating between inputs T and B and output W. In both (**a**) and (**b**) *dashed arrows* indicate constitutive relevance relationships and *continuous arrows* stand for direct causal connections. *Dashed arrows* formally behave exactly like *continuous arrows*

If we want the arrows in our top level model M_2 to still represent the mechanism mediating between input T and B and output W, on the other hand, we can add M only to the micro level structure in M_1 (see Fig. 6.17b).

In both lower level models M_1 in Fig. 6.17 the dashed arrows stand for constitutive relevance relationships, while the continuous arrows still stand for direct causal connections. Both kinds of arrows are intended to work just like ordinary arrows of a BN. This might seem counterintuitive at first glance. Is constitution not a relation that is in many respects totally different from causation? And why should we draw the arrow from the constituent to the constitutor and not the other way round? I will answer these questions by arguing that constitution actually behaves like a special form of causation in Sect. 6.5. For here and now, let me only briefly mention one important aspect causation and constitution do seem to have in common. Both seem to be some kind of grounding relation. Causes as well as constituents seem to determine their effects and constitutors, respectively, in some sense. The effect occurs due to the cause and the constitutor has the properties it has due to the properties of its constituents.

Since I am more interested in mechanisms as devices playing a causal role in mediating between their input and output in this book, I will stick to the kind of representation of mechanisms by variables used in Fig. 6.17b. This kind of representation can be used for diachronical as well as for etiological explanation. It is, hence, clearly the more flexible of the two kinds of representation. Since all the causal work in mediating between the inputs T and B and the output W is done by the mechanism's micro variables (see Sect. 6.5 for more details), it will also do no harm to have M only in the lower level causal model M_1. So M_2 represents the mechanism mediating between inputs T and B and output O at the macro level. The model M_1 tells us what is going on inside the arrows representing this mechanism. In particular, it tells us how the mechanism's micro structure is connected to its input and output variables. If one likes to, one can also add the variable M describing the behavior of the mechanism as a whole.

6.4.4   Philosophical Consequences

In this subsection I want to discuss some of the philosophical consequences of the possibility of modeling mechanisms by means of MLCMs. First of all, let me say that it seems to be quite plausible that providing an MLCM is sufficient for providing an adequate model for mechanistic explanation, prediction, reasoning, and answering questions about how mechanisms would react to certain manipulations. Our representation is as close as possible to the presentation of mechanisms as input-output functions, which is the standard view in the literature. At the macro level, we describe the mechanism's behavior exactly as such an input-output function. When going to the next level down, we deliver detailed information about the mechanism's causal micro structure and how certain inputs causally produce certain outputs, and so on. This seems to be exactly what researchers such as, for example, Craver (2007b) and Woodward (2002, 2013) have in mind. In addition, as we have seen in the last few paragraphs of the last subsection, MLCMs can also be used for diachronical (or constitutional) explanation, prediction, etc., if one wants to do that. From a formal point of view, there really seems to be no difference between the two kinds of explanation/prediction by means of MLCMs. (For more details, see Sect. 6.5.)

However, if my suggestion to model mechanisms by means of MLCMs is adequate, then providing mechanistic explanations seems to give us not much over and above to what providing causal explanations gives us. We begin with the research question how certain inputs to a mechanism (i.e., a certain device at a specific region in space and time) causally produce certain outputs, i.e., with a very coarse-grained causal model. The answer to this question is then given by providing more detailed causal structures connecting the input and output variables and explaining how certain inputs to the mechanism produce its outputs over certain causal pathways through the mechanism's micro structure. So one could say that everything that can be achieved by mechanistic explanation can also be achieved by causal models and the procedure of marginalizing out variables as described in Sect. 6.4.1.

A friend of the new mechanists' approach to explanation may object that we have totally underestimated one important thing about mechanisms, *viz.* constitutive relevance relations. So some may say that mechanistic explanation does not only require causal information, but also information about which parts of the mechanism of interest are constitutively relevant for the phenomenon to be explained; for providing mechanistic explanation it is not sufficient to just provide some causal micro structure, we have to provide the causal micro structure featuring all and only the constitutively relevant parts of the mechanism of interest. The importance of such constitutive relevance relations can be seen, so they may continue, by observing that scientists make top-down as well as bottom-up experiments to discover such relationships. They only claim to be able to explain a certain phenomenon when they have detected the parts of a mechanism that are constitutively relevant for the behavior of interest.

My answer to this objection is the following: First of all, constitutive relevance can be represented in MLCMs, if one wants to do that. (See the end of Sect. 6.4.3.) However, there is no commonly accepted method for uncovering constitutive relationships. (For some serious problems the so far "best-developed account for [mechanistic explanation]" Fagan 2013, p. 100 and for uncovering constitutive relevance relations has to face see, for example, Baumgartner and Gebharter 2016; Leuridan 2012.) It is not even clear so far if such constitutional relations can be empirically detected at all, i.e., uncovered by means of observation and experimentation. (I will sketch a novel suggestion how constitutive relevance relations might be uncovered and distinguished from causal relations in Sect. 6.5.) But more importantly: It is not clear what knowledge about constitutively relevant parts should give us over and above what we get by providing a causal model that represents the system of interest's underlying causal structure in detail. There is no reason why our predictions should become more accurate, nor why our explanations should become better (in whatever sense) when we add information about constitutive relevance relations to our models. Adding constitutive relevance information amounts to adding M and the arrows connected to M to the water dispenser MLCM as in Fig. 6.17b. If we would marginalize out M, however, we would get a model that allows for exactly the same explanations and predictions. Actually, an analysis of constitutive relevance within a causal Bayes nets framework shows that all the causal work is done by the constituents and their causal connections to each other as well as to the input and output variables of the mechanism. (For more details, see Sect. 6.5.)

It is also not clear why information about which parts are the constitutively relevant ones should give us any additional knowledge about manipulation and control. In addition, it is not clear why the causal micro structure used to mechanistically explain a certain phenomenon should contain all and only the constitutively relevant parts of the mechanism. As in standard causal explanation, it may as well be nice to know just *some* causally relevant factors of the phenomenon to be explained. Of course, the more causal details we learn, the better our explanations and the more accurate our predictions will become. But full causal knowledge is neither necessary for causal explanation, nor for prediction. Why should full knowledge about a mechanism's constitutively relevant parts be necessary for mechanistic explanation? And also, why should it be problematic if the causal micro structure of a model used for giving a mechanistic explanation contains some variables not representing constitutively relevant parts of the mechanism? Having such variables in ones model will do no harm to a given explanation/prediction; at least I can see no reason to think otherwise.

Here is another thing about constitutive relationships and scientific practice. Granted, scientists carry out top-down as well as bottom-up experiments. But I am not convinced that the main purpose of these experiments is to learn something about constitutive relevance. Contrary to some philosophers of science, scientists do typically not speak of constitutive relevance relations at all. Top-down and bottom-up experiments do also make sense without the goal of uncovering constitutive relevance relationships. They do also allow to uncover the causal micro structure of a mechanism which connects this mechanism's input variables to its output variables. If I manipulate a mechanism's input variables at the macro level and observe that this leads to a change of the mechanism's output, then I can learn something about the mechanism's causal micro structure: I just have to look at the mechanism's micro structure and check which of the variables I use to model this structure change. These variables may then be good candidates for giving a causal explanation of the mechanism's output. And vice versa, by manipulating some of the mechanism's micro variables, I can check which of these micro variables lead to a change in the mechanism's output. If these variables can also be influenced by the mechanism's input variables of interest, then they are good candidates for explaining the phenomenon I want to explain. Discovering such causal micro structures gives me what scientists want to achieve: knowledge that allows to causally explain the phenomena of interest, that allows for prediction of these phenomena, and that provides answers to questions about what would happen if I would intervene in the system of interest in certain ways. Everything needed for this seems to be delivered by MLCMs. So from an empirical point of view, is seems that information about constitutive relevance might be more or less superfluous. This will become even more clear in the next section.

6.5 Constitutive Relevance Relations in MLCMs

6.5.1 Constitutive Relevance Relations in Mechanisms

If one is interested in diachronic mechanistic explanation, then one is typically also interested in constitutive relevance. One wants to distinguish the parts of the system of interest which are relevant for a certain behavior of interest from the irrelevant parts. It is, however, still an open problem how constitutive relevance can be accounted for and how it can be defined. The most prominent approach to capture constitutive relevance at the market is probably Craver's (2007a, b) mutual manipulability approach.[8] In a nutshell, Craver's idea is that the behavior of a part u of a system s of interest is relevant for a certain behavior of s if and only if there is an intervention on u's behavior that is associated with s's behavior and, vice versa, there is an intervention on s's behavior that is associated with

---

[8] An alternative approach based on a regularity theory of constitution is, for example, developed by Harbecke (2015).

u's behavior. Though Craver introduces interventions in a way slightly different from Woodward (2003), Craver interventions are, strictly speaking, nothing over and above Woodwardian interventions (see also Sect. 5.2).[9] However, it turned out that the mutual manipulability criterion combined with a Woodwardian interventionist theory of causation leads to (more or less) absurd consequences. Leuridan (2012) has shown that within Woodward's interventionist framework the mutual manipulability approach leads to the consequence that constitution is a special form of bi-directed causation. This is a consequence most mechanists would like to avoid (cf. Craver and Bechtel 2007). If one assumes that constitutive relevance is a non-causal relation, on the other hand, then it turns out that there cannot be any constitutive relevance relation in any mechanism whatsoever (Baumgartner and Gebharter 2016), which seems to be even more absurd. In the latter paper we also suggest a modification of an interventionist theory that would actually be compatible with the mutual manipulability criterion. But also within this modified theory mutual manipulability would still not give clear evidence for constitutive relevance relations. In principle, the mutual manipulability of macro and micro phenomena within this modified theory could also arise due to common causes of these phenomena. We provided an additional criterion, the fat-handedness criterion, which gives—together with mutual manipulability—at least some kind of abductive support for constitutive relevance relations. This project of an abductive method to infer constitutive relevance relations has been further developed by Baumgartner and Casini (in press). In another work (Gebharter 2015) I found, however, that interventionist theories such as Woodward's and also the modification we used in Baumgartner and Gebharter (2016) have some more general problems when dealing with systems featuring relationships of constitution, supervenience, and the like. I cannot go into the details here and rather content myself with the following general remark (for more details, see the mentioned papers): It seems that interventionist theories do have a blind spot when applied to variables standing in non-causal relationships such as constitution and supervenience. Such theories sometimes output causal arrows without any evidence. It seems, thus, to be promising to search for a technique for identifying constitutive relevance relations that does not require interventions in the sense of Woodward.

This is basically what I will do in Sect. 6.5. But before I can do that, I have to say a little bit more about constitutive relevance in mechanisms. I will bracket all controversies surrounding this relation and only focus on what is commonly accepted. In particular, I will stay neutral on whether mechanists are obligated to some form of ontological reductionism (cf. Eronen 2011; Fazekas and Kertesz 2011; Kistler 2009; Soom 2012) and on whether constitutive relevance is some kind of causal relation. It might, however, become more plausible to regard constitutive relevance as a causal relation in the light of my results. But I do not want to assume that constitutive relevance is a causal relation in advance. I am also not aiming

---

[9]For a proof that Craver interventions actually are Woodwardian interventions, see Leuridan (2012).

at providing a definition of constitutive relevance. I prefer to treat constitutive relevance and causation in the same way as theoretical concepts are treated in empirical theories. In Glymour's (2004) words, I prefer an Euclidean approach to constitutive relevance over a Sokratic one. I will only provide some characteristic marks of the constitutive relevance relation and then see what we can learn on that basis about constitutive relevance.

Here are the uncontroversial characteristic marks I will use later on: First of all, constitutive relevance is a relation between the behavior of certain parts of a system and certain behaviors of that system as a whole. I will, however, model constitutive relevance at a more abstract level. I will be mainly interested in constitutive relevance between variables. Here is how constitutive relevance between variables is connected to constitutive relevance between behaviors of parts and wholes: Let $X_1, \ldots, X_n$ be variables whose values describe the possible behaviors of the parts of a system. Let the system's possible behaviors be the values of a variable Y. Now X_i is constitutively relevant for Y if some behaviors x_i are constitutively relevant for some behaviors y. And vice versa: When X_i is constitutively relevant for Y, then there will be some X-values x that are constitutively relevant for some behaviors y of Y.

The second characteristic mark is that the behaviors of constitutively relevant parts of a system can be expected to regularly change simultaneously with the behavior of the system as a whole. This is one of the main differences to ordinary causal relations. In case of ordinary causal relations the effect is expected to always occur a little bit after the cause has occurred.

The third characteristic mark is that the behavior of the system as a whole supervenes on the behaviors of its parts. This means that every change in our macro variable Y's value has to go hand in hand with a change in the value of at least one of the micro variables X_i. This also implies that changes in Y-values have to be associated with changes of the probability distribution over the micro variables $X_1, \ldots, X_n$. (We assume here that $X_1, \ldots, X_n$ describe all of the mechanism's constitutively relevant parts.)

$$\forall y, y' \exists x_1, \ldots, x_n : y \neq y' \Rightarrow P(x_1, \ldots, x_n | y) \neq P(x_1, \ldots, x_n | y')$$

While the behavior of the macro variable in a mechanism supervenes on the behaviors of its constitutively relevant parts, the behaviors of all the relevant parts together constitute (and determine) the state of the system as a whole. This means that we can predict the behavior of the system as a whole with probability 1 once we have learned how its constitutively relevant parts behave. This implies the following constraint on our probability distribution P:

$$\forall x_1, \ldots, x_n \exists y : P(y | x_1, \ldots, x_n) = 1$$

The four characteristic marks mentioned above are typically assumed to be met by all mechanisms. There is also a fifth characteristic mark that can be expected to hold for almost all mechanisms: multiple realizability. There are, typically, many ways

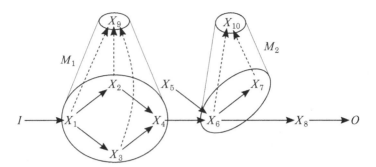

Fig. 6.18 Exemplary system featuring two mechanisms

a mechanism's macro behavior can be realized by its micro parts' behaviors. Also this characteristic mark gives us a constraint on our probability distribution P. If a mechanism's macro behaviors are multiple realizable, then the probabilities for all possible instantiations of the system's micro variables will lie somewhere in the interval $[0, 1[$:

$$\forall x_1, \ldots, x_n, y : P(x_1, \ldots, x_n | y) < 1$$

Throughout Sect. 6.5 I will use the following abstract toy example (see Fig. 6.18). The variables of interest are $X_1, \ldots, X_{10}$. I stands for causal input and O for the causal output of the system we are interested in. In our system, we have two mechanisms: X_9 describes the possible behaviors of the first mechanism M_1 at the macro level, X_{10} describes the possible behaviors of the second mechanism M_2 at the macro level. The variables in the ellipses connected to X_9 and X_{10} describe the behaviors of some of these mechanism's parts. So $X_1, \ldots, X_4$ describe behaviors of parts of the first mechanism at the micro level, and X_6 and X_7 describe behaviors of some of the parts of the second mechanism. Continuous arrows stand for direct causal connections, while dashed arrows stand for constitutive relevance relations. Note that not all of the first mechanism's micro variables are constitutively relevant for X_9. Only X_1, X_2, and X_3 are assumed to be constitutively relevant. We also assume that the characteristic marks of mechanisms introduced above are satisfied by the two mechanisms in our exemplary system. Hence, if all constitutively relevant variables for X_9 and all constitutively relevant variables of X_{10} are captured by $V = \{X_1, \ldots, X_{10}\}$, then due to supervenience there will be values $x_1, \ldots, x_3$ for every value x_9 such that $P(x_1, \ldots, x_3 | x_9) \neq P(x_1, \ldots, x_3 | x_9')$ holds, and there will be values x_6 and x_7 for every value x_{10} such that $P(x_6, x_7 | x_{10}) \neq P(x_6, x_7 | x_{10}')$ holds. Due to constitution there will be an X_9-value x_9 for every combination of values $x_1, \ldots, x_3$ of the micro variables $X_1, \ldots, X_3$ such that $P(x_9 | x_1, \ldots, x_3) = 1$, and there will be an X_{10}-value x_{10} for every combination of values x_6, x_7 of the micro variables X_6, X_7 such that $P(x_{10} | x_6, x_7) = 1$. Finally, due to multiple realizability the probabilities $P(x_1, \ldots, x_3 | x_9)$ for every combination of $X_1, \ldots, X_3, X_9$-values $x_1, \ldots, x_3, x_9$ will lie somewhere within the interval $[0, 1[$, and also the probabilities $P(x_6, x_7 | x_{10})$ for every combination of X_6, X_7, X_{10}-values x_6, x_7, x_{10} will lie somewhere within the interval $[0, 1[$.

Finally, one may wonder why there are no causal arrows entering or exiting one of the macro variables X_9 and X_{10}. I directly causes one of X_9's constitutively relevant variables. So why is I not also a direct cause of the mechanism M_1's macro behavior X_9? There are directed causal chains from M_1's constitutively relevant micro variables to M_2's constitutively relevant micro variables X_6 and X_7. But does this not give us evidence that also the mechanism (as a whole) X_9 should be causally relevant for X_6, X_7, and also for M_2's macro behavior modeled by X_{10}? In addition, should X_{10} not be causally relevant for X_6's direct effect X_8? I will come back to these questions later on in Sect. 6.5.3. We will then find that under some circumstances there are good reasons to not draw the mentioned additional causal arrows. For the moment, we can just assume that we are only interested in constitutive relevance and the causal relations between micro variables. This is also exactly what is required for diachronic mechanistic explanation.

6.5.2 Constitutive Relevance and MLCMs

To see how constitutive relevance relations can be implemented in MLCMs, we have first to take a closer look at how causal arrows work in causal models. Bayesian networks and causal models are so nice for representing causal structures because they capture one essential formal property direct causal relations have. Under certain circumstances (see Sect. 3.3 for details) causal models satisfy the causal Markov condition. In such circumstances, direct causal relations produce the Markov factorization:

$$P(x_1, \ldots, x_n) = \prod_i P(x_i | par(X_i)) \tag{6.7}$$

When doing causal modeling, then one of the typical assumptions made is that the variables of interest do not stand in other than causal relations. This guarantees that our models' arrows '$\longrightarrow$' can only represent direct causal relations. Assume, for example, we are interested in $V = \{X, Y\}$ and our system satisfies CMC. We also measure a dependence between X and Y. Then we can conclude that either $X \longrightarrow Y$ or $Y \longrightarrow X$ is the system's causal structure. This conclusion, however, requires the assumption that no non-causal relations are present. To illustrate this, assume now that this assumption is not met and that Y is constituted by X. Y might, for example, describe the form of a statue, while X describes the position of the statue's left arm. In that case, CMC tells us, again, that either X is a direct cause of Y or that Y is a direct cause of X. But this time, both possible consequences are false (assuming that constitution is a non-causal relation). X and Y are not correlated because one causes the other, but because X is one of Y's constituents. In the next paragraphs I want to argue that constitutive relevance shares an important formal property with direct causal relations. Just like direct causal relations, constitutive relevance relations produce the Markov factorization. As a consequence, we can represent constitutive relevance relations in exactly the same way as direct causal relations in causal models.

Here is my argumentation why constitutive relevance relations behave like direct causal relations: We can observe the same screening off properties in case of constitutional chains as in causal chains. Assume, for example, that X constitutes Y and that Y constitutes Z. Assume further that X and Z do not stand in other relationships to each other. Then fixing Y's value will screen X and Z off each other. Once we have learned Y's value, it does not matter anymore for the probability distribution over Z how exactly Y is realized, i.e., which value X has taken. Conditionalizing on X will not give us any additional information in that case. This is exactly the same screening off behavior we would find in case of a causal chain $X \longrightarrow Y \longrightarrow Z$.

Now let us have a look at common constitutor structures. Assume X is constitutively relevant for both Y and Z.[10] Let us further assume that Y and Z are not connected otherwise by a chain of causal or non-causal relations. Then Y and Z can be expected to become independent when conditionalizing on X. If we know X's value, then learning Y's value cannot give us any additional information about the probability distribution over Z, and vice versa: Learning Z's value cannot give us any additional information about the probability distribution over Y. This is, again, exactly the same screening off behavior we would expect in case of a common cause structure $Y \longleftarrow X \longrightarrow Z$.

Let us have a brief look at a constitutive relevance structure in which we have two constitutively relevant variables. Assume, for example, that X and Y are both constitutively relevant for a macro variable Z. Assume, in addition, that X and Y are not connected by other relations. Also here constitutive relevance relations behave exactly like direct causal relations in common effect structures. X and Y are independent unconditionally, but they can (and in most cases will) become dependent when one conditionalizes on Z. Here is a simple example to illustrate this: Assume we are interested in the behavior of a bidder in an auction. Z describes whether the bidder bids or not ($Z = 1$ and $Z = 0$). Bidding is constituted by whether the bidder raises one of her arms. Now X could stand for whether the bidder raises her left arm or not ($X = 1$ and $X = 0$), and Y could describe whether she raises her right arm or not ($Y = 1$ and $Y = 0$). Now our bidder bids ($Z = 1$) if she raises at least one of her arms ($X = 1$ or $Y = 1$), and she does not bid if she does not raise one of her arms ($X = 0$ and $Y = 0$). Note that X and Y are independent. Whether our bidder raises her left or her right arm does only depend on her free decision. However, if we conditionalize on $Z = 1$, then X

---

[10]Y and Z are assumed to describe macro behaviors of different systems in the sense that the supervenience bases of Y and Z are not identical; note that Y and Z's supervenience bases have, of course, to overlab simply because X is constitutively relevant for both Y and Z.

and Y become dependent. If we know that our bidder bids ($Z = 1$), then we also know that she must have raised one of her arms. We do, however, not know which arm she raised. Hence, $P(X = 1|Z = 1)$ as well as $P(Y = 1|Z = 1)$ will be lower than 1. Now assume that we learn that our bidder did not raise her left arm ($X = 0$). Then, of course, she must have raised her right arm ($Y = 1$). Hence, we get $P(Y = 1|X = 0, Z = 1) = 1 > P(Y = 1|Z = 1)$ and X and Y turn out to be linked up by Z. This is exactly the same linking-up behavior we also find in faithful common effect structures $X \longrightarrow Z \longleftarrow Y$.

Let us now ask how our toy system depicted in Fig. 6.18 can be represented within an MLCM. Here is my suggestion (see Fig. 6.19): The lower level causal model contains the full causal and constitutional structure between the input I and the output O. We can use this structure for mechanistic explanation and prediction. We can explain or predict how certain inputs produce certain outputs of the system. This would be the etiological explanation/prediction. But we can also—if we want to—explain or predict the two mechanisms' macro behaviors modeled by X_9 and X_{10}. This would be the diachronic explanation/prediction. On the higher level, we just marginalize out all the variables describing the two mechanisms (at any level). Here causal arrows represent these mechanisms mediating between their respective inputs and outputs. $I \longrightarrow X_4$ represents mechanism M_1 mediating between its input I and its output X_4 (as well as X_8 and O); $X_4 \longrightarrow X_8 \longleftarrow X_5$ represents mechanism M_2 mediating between its inputs X_4 and X_5 (and I) and its output X_8 (and also O). By going from the higher level model to the lower level model we are zooming into the arrows representing the mechanisms M_1 and M_2. The lower level model provides

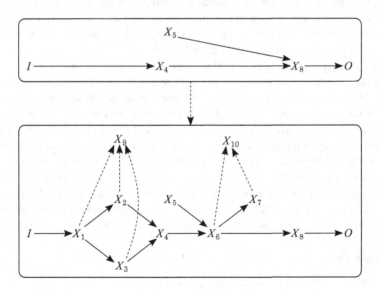

Fig. 6.19 MLCM representation of the exemplary system introduced earlier

detailed information about the relevant parts of mechanisms M_1 and M_2, how they causally interact with each other, and how they are connected to their respective input and output variables.

The suggested representation gives us even something Casini (2016) requested from MLCMs: We can now decide which arrows of the higher level model represent mechanisms just by looking at the MLCM's models' graphs. The bunches of arrows at the higher level we get by marginalizing out clusters of variables connected by paths only consisting of dashed edges represent mechanisms. In particular, we get the arrow $I \longrightarrow X_4$ representing M_1 by marginalizing out $X_1, \ldots, X_3, X_9$, and we get $X_4 \longrightarrow X_8 \longleftarrow X_5$ by marginalizing out X_6, X_7, X_{10}.

In the next subsection I will try to sketch how standard procedures for causal search could be used to uncover causal relations between micro variables as well as constitutive relevance relations between micro and macro variables of a mechanism.

6.5.3   Uncovering Constitutive Relevance Relations

There is a multitude of search algorithms for causal structures available in the literature (see, e.g., Spirtes et al. 2000). All of these algorithms require a measured probability distribution as well as certain assumptions to be satisfied. They then output a structure which may consist of different kinds of edges between variables which stand for different kinds of causal information. If my earlier argumentation is correct and constitutive relevance relations behave like direct causal relations in causal models, then standard search algorithms for causal structures may also be able to output information about constitutive relevance relations. The edges of structures resulting from such algorithms if applied to mechanisms would then either stand for causal or constitutional relations.

How exactly already available algorithms can be used to discover constitutive relevance relations in mechanisms and how relationships of constitutive relevance can be distinguished from causal relations when using which algorithm etc. would be a whole new research project. Here I only want to make the point that such algorithms might be used for such purposes. I will illustrate this by applying the probably best known causal search algorithm to our exemplary system introduced earlier, the PC algorithm (Spirtes et al. 2000, pp. 84f). In what follows $X - Y$ stands short for either $X \longrightarrow Y$ or $Y \longrightarrow X$. $Adj(G, X)$ stands for the set of variables which are adjacent to X in a graph G. These are the variables Y such that $X \longrightarrow Y$, $Y \longrightarrow X$, or $X - Y$ in G. When going through the single steps of the PC algorithm, the graph G as well as the sets $Adj(G, X)$ and $Sep(X, Y)$ are constantly updated.

PC algorithm

S1: Form the complete undirected graph G over the vertex set V.

S2: $n = 0$.

  repeat

repeat

> select an ordered pair of variables X and Y that are adjacent in G such that $Adj(G, X)\backslash\{Y\}$ has cardinality greater than or equal to n, and a subset S of $Adj(G, X)\backslash\{Y\}$ of cardinality n, and if X and Y are independent conditional on S delete edge $X - Y$ from G and record S in $Sep(X, Y)$ and $Sep(Y, X)$;
>
> until all ordered pairs of adjacent variables X and Y such that $Adj(G, X)\backslash\{Y\}$ has cardinality greater than or equal to n and all subsets S of $Adj(G, X)\backslash\{Y\}$ of cardinality n have been tested for independence;
>
> $n = n + 1$;

until for each ordered pair of adjacent vertices X and Y, $Adj(G, X)\backslash\{Y\}$ is of less than n.

S3: For each triple of vertices X, Y, Z such that the pair X, Y and the pair Y, Z are each adjacent in G but the pair X, Z are not adjacent in G, orient $X - Y - Z$ as $X \longrightarrow Y \longleftarrow Z$ iff Y is not in $Sep(X, Z)$.

S4: repeat

> if $X \longrightarrow Y$, Y and Z are adjacent, X and Z are not adjacent, and there is no arrowhead at Y, then orient $Y - Z$ as $Y \longrightarrow Z$;
> if there is a directed path from X to Y, and an edge between X and Y, then orient $X - Y$ as $X \longrightarrow Y$;

until no more edges can be oriented.

PC requires a list of dependence and independence relations that hold in a given measured probability distribution. It also requires CMC, CFC, and that the structures we want to learn do not feature causal cycles. Let us—just for the moment—assume that all of these assumptions are satisfied for our exemplary system. In that case we can read off the dependencies and independencies featured by our exemplary system's probability distribution from the graph depicted in Fig. 6.18. If we use this list as PC's input, then the four steps of PC will lead to the structures depicted in Fig. 6.20. (a) is the result of step 1. In this step we just draw an undirected edge between every two variables in V. The rational behind step 1 is basically to consider all possible causal hypotheses. The result of step 2 is depicted in (b). Here the possible causal hypotheses are thinned out. Because we have assumed that CFC is satisfied, variables that are adjacent in the true causal structure cannot be screened off. So, vice versa, whenever two variables X and Y can be screened off each other by some subset of $V\backslash\{X, Y\}$, they cannot be adjacent and we can delete the undirected edge. Step 3 leads to the structure in (c). Here we can orient some of the undirected edges. Assume X and Y as well as Y and Z are adjacent, but X and Z are not adjacent in the structure resulting from step 2. Then there are three possibilities. Either X and Z are connected by a directed causal path going through Y, Y is a common cause of X and Z, or Y is a common effect of X and Z. If one of the former possibilities would be the case, then there has to be some subset of V containing Y that screens X and Z off each other. And, vice versa, if there is no such screening off set, then

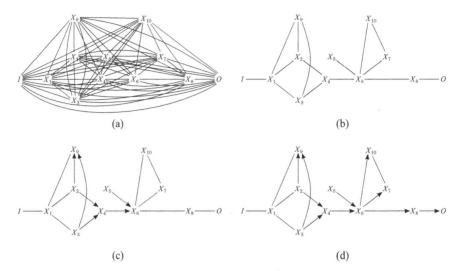

Fig. 6.20 Results of running PC. (**a**) results from step 1, (**b**) from step 2, (**c**) from step 3, and (**d**) from step 4

the only possibility left is that Y is a common effect of X and Z. Step 4 results in the structure depicted in (d). If step 3 already found substructures $X \longrightarrow Y - Z$ (where X and Z are not adjacent), we can orient $Y - Z$ as $Y \longrightarrow Z$. (If Z would have been a direct cause of Y, then step 3 would have told us that.) In addition: If there is a directed causal chain from X to Y and also an undirected edge, then this undirected edge must be directed as $X \longrightarrow Y$. Otherwise the initial assumption of acyclicity would be violated.

In case of our exemplary system, PC allows us to uncover the skeleton of the correct causal and constitutive relevance structure. It provides even information about how most of the edges involved in that structure have to be oriented. PC is also nicely compatible with additional causal background knowledge, which might often be available when we are interested in constitutive relevance discovery in mechanisms. Let us, for example, assume we already know (for what reasons ever) that I is a direct cause of X_1. In that case, we can just add the arrow $I \longrightarrow X_1$ to our graph under construction G before we run PC. The single steps of PC would then give us the structures depicted in Fig. 6.21. Again, (a) is the result of step 1, (b) the result of step 2, (c) the result of step 3, and (d) the result of step 4.

If we look at the structure PC outputs, there is obviously a still unsolved problem. PC was able to correctly identify constitutive relevance relations because they formally behave like direct causal connections in causal Bayes nets. This means, on the other hand, that PC cannot distinguish direct causal connections from constitutive relevance relations. So how could we do that after running PC? Here is one possible way to go: We could try to distinguish constitutive relevance relationships from causal connections by means of time order information. In mechanisms, causes typically precede their effects in time, while behaviors

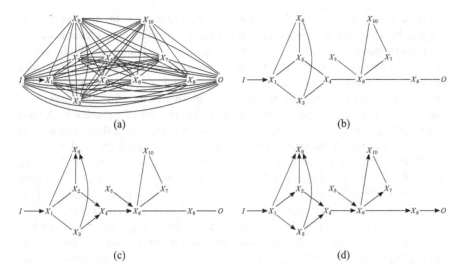

Fig. 6.21 Results from running PC with causal information $I \longrightarrow X_1$. (a) results from step 1, (b) from step 2, (c) from step 3, and (d) from step 4

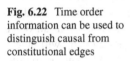

Fig. 6.22 Time order information can be used to distinguish causal from constitutional edges

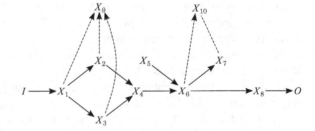

connected by constitutive relevance relations can be expected to regularly occur simultaneously. We could use this difference between the two kinds of relationships to distinguish them in the structure resulting from PC. In particular, we would have to check every pair of variables adjacent in the structure produced by PC for whether they are regularly instantiated together. Since X_9 is assumed to be constituted by X_1, X_2, and X_3, and X_{10} is assumed to be constituted by X_6 and X_7 in our example, we would find that of all the adjacent variables in V only X_9 would regularly be instantiated simultaneously with X_1, X_2, and X_3, while only X_{10} would regularly be instantiated together with X_6 and X_7. Now we can simply replace the continuous edges between these variables by dashed ones. We were finally able to identify and distinguish all causal and constitutive relevance relations in our exemplary system. Figure 6.22 shows the result of applying this last step to the structure output by PC in case we already know that the system's input I is a direct cause of X_1.

Before I will go on and discuss the empirical adequacy of the discovery procedure suggested, I would like to briefly discuss mixed relationships. We have asked how constitutive relevance relations can be distinguished from causal relations in the structure resulting from PC. My claim was that time order information can do the job. Strictly speaking, however, time order information cannot guarantee that a particular continuous arrow stands for a "pure" causal relation. While dashed arrows resulting from our discovery procedure should always stand for constitutive relevance relationships, continuous arrows may either stand for (pure) causal connections or for mixed connections. Let me briefly illustrate this by means of the following example: Assume that we are interested in a set of variables V. Among other variables, V contains X and Z, but not Y. In truth, X is a direct cause of Y, and Y is constitutively relevant for Z. Assume now that we run PC on the variable set V. It might now be the case that PC outputs a structure featuring the arrow $X \longrightarrow Z$, though X is not a cause of Z. X is, in truth, rather a cause of a constitutively relevant variable of Z not in V. So, strictly speaking, the method suggested allows to distinguish constitutive relevance relations from causal *or* mixed relations.

There is another issue deserving attendance before we go on. In the structure in Fig. 6.22 there are no causal arrows exiting one of the macro variables X_9 or X_{10}. This will typically be the case when the respective mechanisms' full supervenience bases are contained in our variable set of interest V. Let us assume that $\{X_1, X_2, X_3\}$ is X_9's full supervenience base and that $\{X_6, X_7\}$ is X_{10}'s full supervenience base. Then, due to the fact that a mechanism's macro behavior is determined by its supervenience base, X_9 is screened off from all other variables in V by its supervenience base $\{X_1, X_2, X_3\}$ and X_{10} is screened off from all other variables in V by its supervenience base $\{X_6, X_7\}$. Hence, step 2 of PC will delete all edges between the two macro variables and variables in V that do not belong to their respective supervenience bases. This has as a side effect that macro variables supervening on micro variables are, in principle, causally inefficacious. (Note that this result only holds for variable sets featuring full supervenience bases.) This observation directly leads to a number of new questions. Here are a few examples: How is it then possible to intervene on macro variables? Does this imply then that causation exists only at the fundamental physical level? Does the whole approach not directly lead into some kind of epiphenomenalism? These are, of course, important questions. There is, however, not enough space in this book to further explore these ontological issues. Here I am more concerned with epistemological questions of how to represent and uncover constitutive relevance relations. For a first glimpse on ontological consequences and on how interventions might work in systems featuring variables standing in supervenience relationships, see Gebharter (2015).

Here is one additional issue also connected with the last one. In case V contains the full supervenience base of one or more variables also in V, then it might seem that faithfulness will regularly be violated.[11] Assume, for example, that we have to deal with a system in which X_1 directly causes X_2, which, in turn, directly causes X_3. Now assume further that X_2's full supervenience base $\{X_4, X_5\}$ is also in V. In that case, so one might object, conditionalizing on $\{X_4, X_5\}$ will screen both X_1 and X_3

---

[11]I am indebted to Michael Baumgartner (personal communication) for that comment.

off from X_2 and faithfulness would be violated. I agree that a structure featuring the causal chain $X_1 \longrightarrow X_2 \longrightarrow X_3$ would not be faithful if X_2's supervenience base is also in V. But why should we agree that this is the correct causal structure in the first place? It seems plausible that in case of mechanisms the whole causal work is done by the micro structure. Since the micro variables constitute the macro variables, what causal role could be left for the macro variables to play? My thoughts on this example are that one should conclude that X_1 is not a direct cause of X_2, but rather of one the two variables X_4 and X_5 constituting X_2. In addition, X_2 should not be modeled as a cause of X_3. Rather, one of its constituents X_4 or X_5 should be a direct cause of X_3.

Of course, during the whole section I just assumed that CMC and CFC are satisfied without any argument why they should be expected to hold in systems featuring causal as well as constitutional relationships. There might be good reasons why violations of one or both of these conditions can be expected in such systems. A thorough investigation of how these two fundamental conditions behave in case of models of mechanisms is something to be carried out in the future.

6.5.4   Empirical Adequacy

During the last subsections I tried to argue that constitutive relevance relations are not that different from ordinary causal relations as many mechanists seem to think. Both kinds of relationships produce the Markov factorization (under suitable conditions). This is everything required for representing them by the arrows of a Bayes net. Actually, Bayes nets are quite flexible; they can represent any kind of relation that produces the Markov factorization. In ordinary causal modeling it is simply assumed that the variable sets of interest do not contain variables standing in other than causal relations. Hence, the edges output by standard search algorithms must stand for certain causal dependencies. If we also allow for constitutive relevance (and, hence, for supervenience) dependencies and if my argumentation that constitutive relevance relations behave like direct causal connections is correct, then the edges output by these search algorithms must stand for causal or constitutive relevance relationships. This result might also make recent suggestions of interpreting constitutive relevance as a special kind of causation (see, e.g., Leuridan 2012) more attractive.

A clear advantage of the discovery procedure suggested would be that it does not rely on interventions, such as Craver's (2007a, b) problematic mutual manipulability approach. However, the problems that come with accounting for constitutive relevance by means of interventions highlighted in Baumgartner and Gebharter (2016) reappear in a slightly different form in a causal Bayes net representation of constitutive relevance relations: In both cases it turns out that it is not possible to intervene on macro variables in the presence of their full supervenience bases.

Here is another advantage of the representation of constitutive relevance relations suggested empirically oriented philosophers might find attractive. The empirical adequacy of the whole approach might be empirically tested. If the approach is correct, then we should not be able to find variables $X_1, \ldots, X_n$ and Y such that our discovery procedure outputs a dashed arrow from Y to one of the variables X_i if $X_1, \ldots, X_n$ describe the behaviors of parts of the system whose behavior as a whole is described by Y (provided CMC and CFC are satisfied). There is also a highly interesting result by Ramsey et al. (2002). Ramsey et al. used a modified version of PC to identify the components of rock and soil samples. They were able to identify the most frequently occurring kinds of carbonates as good or better than human experts. If the constitution relation involved here is similar enough to constitutive relevance in mechanisms, what at least at first glance seems plausible, Ramsey et al.'s findings are strong support for the adequacy of the method for uncovering constitutive relevance in mechanisms I have suggested in this section.

In the next section I will shift the focus from the multi level structuring of mechanisms and constitutive relevance to two other typical characteristic marks of mechanisms. Mechanisms are systems that produce certain phenomena over a period of time. In addition, many mechanisms are self regulatory systems that feature several causal feedback loops. Both characteristic marks are—at least at the moment—still problematic for an MLCM representation of mechanisms.

6.6 Modeling Causal Cycles with MLCMs

One shortcoming that Casini et al.'s (2011) RBN approach and the MLCM approach for modeling mechanisms I presented in Sect. 6.4 have in common is that they only allow for modeling mechanisms which do not feature causal cycles. This is clearly a shortcoming, because many mechanisms (especially in the life sciences) are self regulatory systems featuring feedback loops. But meanwhile Clarke et al. (2014) have further developed Casini et al.'s modeling approach in a follow up paper. While Clarke et al.'s expansion of the RBN approach cannot avoid the three problems discussed in Sect. 6.3, it seems to be able to handle causal feedback. My hope is that also the MLCM approach can learn something from Clarke et al.'s strategy for modeling feedback mechanisms.

When it comes to modeling causal cycles, Clarke et al. (2014) distinguish between two different kinds of problems one may have to face. So-called *static problems* are "situations in which a specific cycle reaches equilibrium [...] and where the equilibrium itself is of interest, rather than the process of reaching equilibrium" (Clarke et al. 2014, sec. 6). A *dynamic problem*, on the other hand, is a "situation in which it is the change in the values of variables over time that is of interest" (ibid.). For each kind of problem Clarke et al. propose a solution: Static problems should be solved by modeling the mechanism of interest by means of (static) cyclic causal models. In that case the causal Markov condition has to be replaced by Pearl's (2000, sec. 1.2.3) d-separation criterion. If, on the other hand, one is interested in how a system behaves over a period of time, one should rather use dynamic Bayesian networks (cf. Murphy 2002) for modeling the mechanism of interest. In that case variables are time indexed and causal cycles are "rolled out" over time.

In this section I will show that similar strategies can be endorsed to model causal feedback by means of MLCMs. The section is based on Gebharter and Schurz (2016). It is structured as follows: In Sect. 6.6.1 I present a simple toy mechanism featuring one causal cycle. In Sect. 6.6.2 I illustrate (by means of this exemplary mechanism) how the static problem can be solved, in Sect. 6.6.3 I show how the dynamic problem can be solved. I compare my solution of both problems within the MLCM approach with Clarke et al.'s (2014) solution within the RBN approach in Sect. 6.6.4. It will turn out that the arrows in the graphs Clarke et al. use for probabilistic prediction in their approach do sometimes not represent the true direct causal relationships. Hence, their models can only predict post observation distributions, but not post intervention distributions. MLCMs, on the other hand, can also be used for predicting post intervention distributions.

6.6.1 A Cyclic Toy Mechanism

In the next subsections I will use the following self regulatory toy mechanism for illustration: a simple temperature regulation system. The mechanism's top level is described by two variables and one causal arrow: OT models the outside temperature, IT stands for the inside temperature, and CK is a variable whose values represent the possible adjustments of a control knob. The outside temperature (OT) and the control knob (CK) are directly causally relevant for the inside temperature (IT). Hence, the graph which is the concatenation of $OT \longrightarrow IT$ and $CK \longrightarrow IT$ represents the temperature regulation system (i.e., our mechanism of interest) at work (see Fig. 6.23). Now assume we are interested in a specific phenomenon, *viz.* the phenomenon that the inside temperature is relatively insensitive to the outside temperature when the temperature regulation system is on. In other words, we are interested in the phenomenon that $P(it|ot, CK = on) \approx P(it|CK = on)$ holds for arbitrarily chosen $it \in val(IT)$ and $ot \in val(OT)$.

For mechanistic explanation it is required to tell a story about over which causal paths through the mechanism's micro structure the system's inputs OT and CK produce its output IT. In particular, we have to provide a second and more detailed causal model that allows for a look inside the arrows $OT \longrightarrow IT$ and $CK \longrightarrow IT$. This more detailed model should also provide information about how exactly arbitrary value changes of OT are compensated in such a way that $P(it|ot, CK = on) \approx P(it|CK = on)$ results. That work is clearly done by a causal feedback loop: A temperature sensor S measures the inside temperature IT and

Fig. 6.23 Causal structure of the temperature regulation system at the top level

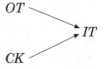

gives input to an air conditioner AC, which, in turn, overwrites the causal influence of the outside temperature OT on the inside temperature IT. So our lower level causal model would have to feature the causal cycle $IT \longrightarrow S \longrightarrow AC \longrightarrow IT$. But recall that, according to our notion of a restriction (Definition 6.3), the causal Markov condition (CMC) is required to hold for the lowest level causal model of the MLCM. As we already saw in Sect. 3.3, the causal Markov condition, though it can in principle be applied to cyclic causal models, does have some serious problems with such models.

The next two subsections show by means of the exemplary mechanism introduced above how the MLCM approach can be modified in such a way that it can handle static as well as dynamic problems. For solving the static problem we have to provide a causal explanation of why $P(it|ot, CK = on) \approx P(it|CK = on)$ holds when the cycle $IT \longrightarrow S \longrightarrow AC \longrightarrow IT$ has reached equilibrium. Solving the dynamic problem requires us to tell a causal story about how the cycle $IT \longrightarrow S \longrightarrow AC \longrightarrow IT$ produces $P(it|ot, CK = on) \approx P(it|CK = on)$ over some period of time.

6.6.2 Solving the Static Problem

To solve the static problem we can simply replace condition (c) of Definition 6.3, which requires the lowest level model of an MLCM to satisfy CMC, by a condition that requires the lowest level model of an MLCM to satisfy the d-connection condition. Recall that the d-connection condition is equivalent to CMC for acyclic causal models (Lauritzen et al. 1990). The main advantage of the d-connection condition over CMC is that it does, contrary to CMC, imply the independence relationships to be expected when applied to causal models featuring cycles (Pearl and Dechter 1996; Spirtes 1995).

With this slight modification of Definition 6.3 the static problem for our exemplary mechanism can be solved by providing the following MLCM (for a graphical illustration, see Fig. 6.24): A causal model M_2 with the graph $OT \longrightarrow IT \longleftarrow CK$ represents the mechanism's top level. Another causal model M_1 describes the mechanism's causal structure at the lower level: The outside temperature (OT) is directly causally relevant for the inside temperature (IT), which is measured by a temperature sensor (S). This sensor as well as the adjustment of the control knob (CK) have a direct causal influence on the state of an air conditioner (AC). The air conditioner, in turn, is directly causally relevant for the inside temperature (IT). Note that M_2 is assumed to be a restriction of M_1 and that M_1 satisfies the d-connection condition.

Now the lower level model M_1 can explain why $P(it|ot, CK = on) \approx P(it|CK = on)$ holds for arbitrary OT- and IT-values when the cycle $IT \longrightarrow S \longrightarrow AC \longrightarrow IT$ has reached equilibrium as follows: If $CK = off$, then also AC will take its off-value. But then there is no self regulation over the feedback loop $IT \longrightarrow S \longrightarrow AC \longrightarrow IT$. Hence, OT's influence on IT will not be canceled. However, if CK takes one of its

Fig. 6.24 MLCM for the
static problem. When
$CK = on$, AC responses to S
and the *bold arrow*
$AC \longrightarrow IT$ overwrites
$OT \longrightarrow IT$

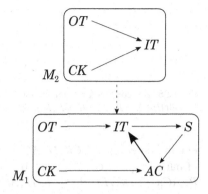

on-values, then AC will respond to S according to CK's adjustment. Now IT's value
is robust to OT-variations when $CK = on$ because the arrow $AC \longrightarrow IT$ overwrites
the causal influence of OT on IT transported over the arrow $OT \longrightarrow IT$ if $CK = on$.

Before I will go on to show how also the dynamic problem can be solved, let
me briefly highlight an open problem: Cyclic causal models featuring bi-directed
arrows do not admit the Markov factorization, meaning that $P(X_1, \ldots, X_n)$ cannot
be computed as the product of the probabilities of each X_i in the ordering conditional
on its causal parents $Par(X_i)$. So it is unclear whether there is a factorization criterion
equivalent to the d-connection condition. We can, however, provide at least a weaker
criterion. Here is my proposal: We can construct a separator set $dSep(X_i)$ that can
more or less play the same role as the set of a variable X_i's causal parents does
in the Markov factorization. We start with an ordering of variables $X_1, \ldots, X_n$. Let
X_i be any arbitrarily chosen variable of this ordering. Next we search for subsets
$dPred(X_i)$ of X_i's predecessors $Pred(X_i)$ in the given ordering for which there are
subsets U satisfying the following conditions: (i) $dPred(X_i) \cup U = Pred(X_i)$, (ii)
$dPred(X_i) \cap U = \emptyset$, and (iii) U d-separates X_i from $dPred(X_i)$. If we cannot find
such subsets $dPred(X_i)$, then we identify $dSep(X_i)$ with $Pred(X_i)$. If, on the other
hand, we can find such sets $dPred(X_i)$, then we choose one of the largest of these
sets and identify $dSep(X_i)$ with this set's corresponding separator set U.

With this method at hand we can construct a set $dSep(X_i)$ for every X_i in a
given ordering $X_1, \ldots, X_n$ that renders X_i independent of all of its predecessors
in the ordering that will oftentimes be much narrower than the set of all of X_i's
predecessors (cf. Gebharter and Schurz 2016, sec. 4.1). The probability distribution
of a cyclic causal model possibly featuring bi-directed arrows then factors according
to the following equation:

$$P(X_1, \ldots, X_n) = \prod_{i=1}^{n} P(X_i | dSep(X_i)) \qquad (6.8)$$

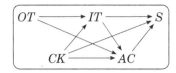

Fig. 6.25 BN constructed from the factorization $P_1(OT, CK, IT, AC, S) = P_1(OT) \cdot P_1(CK) \cdot P_1(IT|OT, CK) \cdot P_1(AC|OT, CK, IT) \cdot P_1(S|CK, IT, AC)$

For the ordering OT, CK, IT, AC, S, for instance, the probability distribution P_1 of our exemplary causal model M_1 factors as $P_1(OT) \cdot P_1(CK) \cdot P_1(IT|OT, CK) \cdot P_1(AC|OT, CK, IT) \cdot P_1(S|CK, IT, AC)$.

The factorization criterion suggested has, however, two disadvantages: The first disadvantage is that it depends on an ordering of the variables of interest. The second disadvantage is that it is weaker than the d-separation criterion, i.e., it may not imply all the independence relations implied by the d-separation criterion. Applied to our exemplary causal model M_1 it does, for instance, not imply the independence $INDEP_{P_1}(OT, S|\{IT, AC\})$, though OT and S are d-separated by $\{IT, AC\}$ in the model's graph. To see this, we can construct a new BN from the factorization $P_1(OT, CK, IT, AC, S) = P_1(OT) \cdot P_1(CK) \cdot P_1(IT|OT, CK) \cdot P_1(AC|OT, CK, IT) \cdot P_1(S|CK, IT, AC)$ in which the parental relationships correspond to the conditional probabilities in the factorization. Thus, this BN's graph would be the one depicted in Fig. 6.25. OT and S are not d-separated by $\{IT, AC\}$ in this graph, and thus, the factorization does not imply the independence $INDEP_{P_1}(OT, S|\{IT, AC\})$ which is implied by the d-separation criterion applied to the original graph of M_1.

6.6.3 Solving the Dynamic Problem

Clarke et al.'s (2014) solution of the dynamic problem consists in adding time indices to the system of interest's variables. When doing this, causal cycles can be "rolled out" over time. Clarke et al. use so-called dynamic Bayes nets (cf. Murphy 2002) for this purpose. Instead of dynamic Bayesian networks I use dynamic causal models (DCMs) which do also allow for bi-directed arrows.

DCMs are quadruples $\langle V, E, P, t \rangle$, where t is a time function that assigns a positive natural number to every variable in V (i.e., $t : V \rightarrow \mathbb{N}^+$). In a DCM V consists of infinitely many variables $X_{1,1}, \ldots, X_{n,1}, X_{1,2}, \ldots, X_{n,2}, \ldots$. Variables $X_{1,t}, \ldots, X_{n,t}$ typically model the behavior of different entities at the same time. The second index (i.e., the t) stands for the time (or stage) at which the respective parts' behaviors are described. So the variables $X_{1,1}, \ldots, X_{n,1}$, for example, describe the system modeled at its initial state (stage 1), $X_{1,2}, \ldots, X_{n,2}$ describe the system at stage 2, etc.

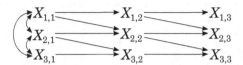

Fig. 6.26 Three stage finite segment of an ideal DCM. The first stage features three bi-directed *arrows* accounting for correlations between $X_{1,1}$ and $X_{2,1}$, $X_{2,1}$ and $X_{3,1}$, and $X_{1,1}$ and $X_{3,1}$ due to past common causes not included in V but of the kind as described by the variables in V

The DCMs we will be interested in all presuppose the following two idealizations:

- If $X_{i,t} \longrightarrow X_{j,t'}$, then $t' > t$, i.e., a directed arrows's head is always at a later stage than its tail.
- If $X_{i,t} \longrightarrow X_{j,t+u}$ holds for some stage t, then $X_{i,t} \longrightarrow X_{j,t+u}$ holds for all stages t, i.e., the pattern of directed arrows between two stages t and $t + u$ spreads to all stages t and $t + u$.

An ideal DCM (see Fig. 6.26 for a finite segment of such a DCM) will, in addition, also have the following four properties:

- If $X_{i,t}$ @—@ $X_{j,t'}$, then $|t - t'| \leq 1$, i.e., arrows do not skip stages.
- If $X_{i,t} \longleftrightarrow X_{j,t'}$, then $t = t'$, i.e., bi-directed arrows connect only variables of one and the same stage.
- For all $t, t' > 1$: If $X_{i,t} \longleftrightarrow X_{j,t}$, then $X_{i,t'} \longleftrightarrow X_{j,t'}$, i.e., all later stages share the same pattern of bi-directed causal arrows.
- $P(X_{i,t}|Par(X_{i,t})) = P(X_{i,t+1}|Par(X_{i,t+1}))$ holds for all $X_{i,t} \in V$ with $t > 1$.

How can we solve the dynamic problem within the MLCM approach by means of DCMs now? Our exemplary toy mechanism can be modeled by an MLCM consisting of two DCMs M_1 and M_2. M_2 is a restriction of M_1 and models the mechanism's top level. M_1 provides more details about the temperature regulation system and is assumed to satisfy the d-connection condition. Figure 6.27 shows the MLCM's level graph, where M_1 and M_2 are represented by finite segments.

Note that for M_2 to be a restriction of M_1 it is required that the two models' probability distributions fit together. This requires that $t^* = t$, $t^* + 1 = t + 1/3$, $t^* + 2 = t + 2/3$, $t^* + 3 = t + 3/3$ etc., where t stands for M_2's and t^* for M_1's stages. Also note that when going from the higher level model M_2 to the lower level model M_1, we will typically not only add additional variables, but also additional stages for each newly added intermediate cause. In case of our temperature regulation mechanism we added two new intermediate causes, viz. S and AC, together with two new stages, arriving at $IT_{t^*} \longrightarrow S_{t^*+1} \longrightarrow AC_{t^*+2}$ in M_1. Now the graph of M_1 gives us detailed information about the system's causal micro structure: The outside temperature (OT) and the inside temperature (IT) are direct causes of themselves at the next stage. Furthermore, the outside temperature (OT) is also a direct cause of the inside temperature (IT) at the next stage, while the inside temperature (IT) also directly causes the sensor reading (S) at the next stage. The sensor (S), in

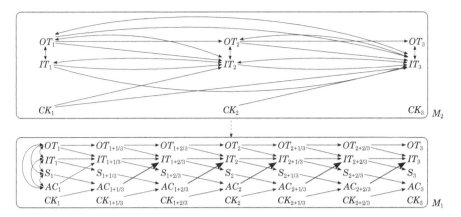

Fig. 6.27 Segments of M_1 and M_2 of an MLCM for the dynamic problem. When $CK_{t^*} = on$ over several stages, AC_{t^*+1} responses to S_{t^*} and *bold arrows* $AC_{t^*+1} \longrightarrow IT_{t^*+2}$ overwrite $OT_{t^*+1} \longrightarrow IT_{t^*+2}$ and $IT_{t^*+1} \longrightarrow IT_{t^*+2}$

turn, is directly causally relevant for the behavior of the air conditioner (AC) at the next stage. Also the control knob's adjustment (CK) is directly causally relevant for the air conditioner (AC) at the next stage. As in the abstract example in Fig. 6.26, the bi-directed arrows between variables at stage 1 indicate common causes not represented in the DCM.

Note that M_1 satisfies all the conditions listed earlier for ideal DCMs. Unfortunately, model M_2 is not that nice. Since we marginalized out S and AC and there were directed paths from OT_{t^*} to IT_{t^*+6} and from IT_{t^*} to IT_{t^*+6} all going through S_{t^*+3} or AC_{t^*+3} in M_1, M_2 features directed arrows $OT_t \longrightarrow IT_{t+2}$ and $IT_t \longrightarrow IT_{t+2}$ skipping stages. Since there were paths indicating a common cause path between OT_{t^*} and IT_{t^*+6} going through S_{t^*+3} or AC_{t^*+3} in M_1, M_2 features bi-directed arrows $OT_t \longleftrightarrow IT_{t+2}$. Note that there are also bi-directed arrows between OT_t and IT_{t+1} and between IT_t and IT_{t+1}.

Now the MLCM mechanistically explains why IT is relatively robust w.r.t. OT-changes when $CK = on$ over a period of time. If CK is *off* over several stages, then also AC is *off* and there is no regulation of IT over paths $IT_{t^*} \longrightarrow S_{t^*+1} \longrightarrow AC_{t^*+2} \longrightarrow IT_{t^*+3}$; IT's value will increase and decrease (with a slight time lag) with OT's value. If, however, CK is fixed to one of its *on*-values over several stages, then over several stages AC_{t^*+1} responses to S_{t^*} according to CK_{t^*}'s adjustment. Now the crucial control mechanism consists of IT_{t^*+1} and its parents OT_{t^*}, IT_{t^*}, and AC_{t^*}. $AC_{t^*} \longrightarrow IT_{t^*+1}$ overwrites $OT_{t^*} \longrightarrow IT_{t^*+1}$ and $IT_{t^*} \longrightarrow IT_{t^*+1}$ when $CK_{t^*} = on$, i.e., $P_{CK_{t^*}=on}(it_{t^*+1}|ac_{t^*}, u_{t^*}) \approx P_{CK_{t^*}=on}(it_{t^*+1}|ac_{t^*})$ holds, where $U_{t^*} \subseteq \{OT_{t^*}, IT_{t^*}\}$. This control mechanism will, after a short period of time, cancel deviations of IT's value from CK's adjustment brought about by OT's influence.

Here are some possible open problems: First, note that some of the causal arrows in M_2 may seem to misrepresent the "true" causal processes going on inside the temperature regulations system. There is, for example, a directed arrow going from

CK_t to IT_{t+1}, but no directed arrow from CK_t to AC_{t+1}, though the control knob can clearly influence the inside temperature only through the air conditioner. This is a typical problem arising for dynamic models. One can, however, learn something about M_1's structure from M_2: The (direct or indirect) cause-effect relationships among variables in M_2 will also hold for M_1. Another problem is, again, search. For solutions of several discovery problems involving time series, see, for example, Danks and Plis (2015); these authors do also present a simple and sparse method for graphically representing ideal DCMs. Finally, factorization and interventions: Since our DCMs do not feature feedback loops, it seems plausible to conjecture that Richardson's (2009) factorization criterion and Zhang's (2008) results about how to compute the effects of interventions in models with bi-directed arrows can be fruitfully applied to DCMs.

6.6.4 MLCMs vs. RBNs Again

In this subsection I compare the suggestions made in Sects. 6.6.2 and 6.6.3 of how to solve static and dynamic problems with Clarke et al.'s (2014) solution. I start with comparing the two solutions of the static problem: Though both approaches use Pearl's (2000) notion of d-separation instead of CMC to account for cycles in case of static problems, the structures used for probabilistic reasoning differ in the two approaches. Clarke et al. use the "true" cyclic graph to construct an equilibrium network, i.e., a Bayesian network that is then used "to model the probability distribution of the equilibrium solution" (Clarke et al. 2014, sec. 6.1). This move seems to have at least two shortcomings, which I will illustrate by means of our exemplary mechanism.

First of all, we have to construct the equilibrium network[12]: We start with our original causal model M_1. As a first step, we search in M_1's graph (depicted in Fig. 6.28a) for substructures of the form $X \longrightarrow Z \longleftarrow Y$. If we find such a substructure and X and Y are not already connected by a causal arrow, we add an undirected edge $X - Y$. The result is the graph depicted in Fig. 6.28b. Next, we replace every directed arrow by an undirected edge $X - Y$. The resulting graph is called the original graph's *moral graph*. It is depicted in Fig. 6.28c. M_1's moral graph contains all the independence information also stored in M_1's original graph: Two variables X and Y are separated by a set of variables Z in the moral graph if and only if they are d-separated by Z in M_1 (cf. Lauritzen et al. 1990).

As a last step, the undirected moral graph has to be transformed into a DAG that, together with the underlying probability distribution, satisfies the Markov condition. The resulting BN is the much sought-after equilibrium network. The equilibrium network can be constructed on the basis of the moral graph in Fig. 6.28c as follows:

---

[12]For more details how to construct equilibrium networks, see Clarke et al. (2014, sec. 6.1, appendix).

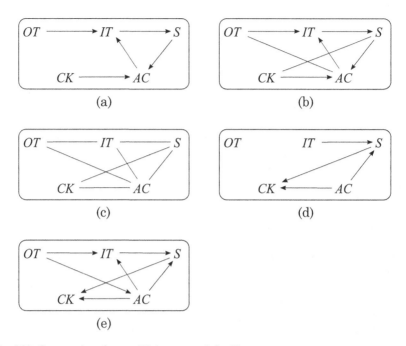

Fig. 6.28 Construction of an equilibrium network for M_1

First, we have to order the variables in V. We arbitrarily select one of the variable of V as our first variable. Let this variable be OT. Next, we choose one of the other variables which has as many connections as possible to the variables which already have been ordered. The two possible candidates are IT and AC, which are both connected to OT by an undirected edge. Let us choose IT. So our ordering so far is OT, IT. We repeat this step: We search for a variable connected to as many variables in $\{OT, IT\}$ as possible. Now there is only one variable connected to both OT and IT, *viz.* AC. So we choose AC and arrive at the ordering OT, IT, AC. Since S is connected to more variables in $\{OT, IT, AC\}$ than CK, S is the next variable in our ordering. We finally arrive at the ordering OT, IT, AC, S, CK.

Next we search for the cliques, i.e., the sets consisting of the variables of the maximal complete subgraphs of our moral graph. There are three cliques: $D_1 = \{OT, IT, AC\}$, $D_2 = \{IT, AC, S\}$, and $D_3 = \{AC, S, CK\}$. Now for each clique D_i we construct two subsets E_i and F_i. E_i is the intersection of the union of D_i and all D_j with $j < i$. Thus, $E_1 = \emptyset$, $E_2 = \{IT, AC\}$, and $E_3 = \{AC, S\}$. F_i is constructed as $D_i \backslash E_i$. Hence, $F_1 = \{OT, IT, AC\}$, $F_2 = \{S\}$, and $F_3 = \{CK\}$.

Finally, we construct the equilibrium network as follows: We draw an arrow from every variable in E_i to every variable in F_i. Hence, we draw arrows $IT \longrightarrow S$, $AC \longrightarrow S$, $AC \longrightarrow CK$, and $S \longrightarrow CK$ and arrive at the graph in Fig. 6.28d. As a last step, we have to check for whether all pairs of variables in the cliques D_i are connected by an arrow. If not, then we have to connect them by an arrow such that the resulting graph does not feature cycles. Hence, we have to add an arrow between OT and IT, OT and AC, and IT and AC. If we decide to add the arrows according to the true cause-effect relationships among the variables in V, we add $OT \longrightarrow IT$, $OT \longrightarrow AC$, and $AC \longrightarrow IT$. The result is the equilibrium network in Fig. 6.28e. The equilibrium network will satisfy the Markov condition. All independencies implied by the equilibrium network and the Markov condition will also be implied by our original model M_1 and the d-separation criterion. Hence, the equilibrium network can be used for making predictions based on observation.

Now let me highlight the two shortcomings mentioned above: (i) Though all independencies implied by the Markov condition and the equilibrium network are also implied by M_1 and the d-separation criterion, independencies implied by the d-separation criterion and the original cyclic causal structure may not be implied by the equilibrium network. Let me illustrate this by means of our model M_1, whose equilibrium network could be the one depicted in Fig. 6.28e: Note that OT and CK are not d-separated (by the empty set) in the equilibrium network. So the equilibrium network's graph does not capture the independence between OT and CK (given the empty set) implied by the d-separation criterion and the fact that OT and CK are d-separated (by the empty set) in M_1's graph.

(ii) Since the arrows of the equilibrium network do not capture the "true" causal relations anymore, it cannot be used for predicting the effects of interventions. To illustrate this, assume that we are interested in the post intervention probability $P(s|do(ck))$ in our model M_1. In case we use M_1's graph for computing this probability, we arrive at $P(s|do(ck)) = P(s|ck)$. If we use the equilibrium network's graph, however, we have to delete the arrows $S \longrightarrow CK$ and $AC \longrightarrow CK$ and, hence, arrive at $P(s|do(ck)) = P(s)$.

Clarke et al.'s (2014) solution to the dynamic problem shares shortcoming (ii) with their solution to the static problem. Surprisingly, also their dynamic Bayesian networks are not intended to represent the "true" causal structure (cf. Clarke et al. 2014, sec. 6.2). This is enough for problem (ii) to occur, since problem (ii) just requires that some causal arrows do not represent true cause-effect relations.

6.7 Conclusion

In Chap. 6 I investigated the question in how far the causal nets approach can be used to contribute to the discussion about mechanisms in the current philosophy of science literature. I started with a few words on mechanisms in Sect. 6.1 and proceeded by presenting Casini et al.'s (2011) RBN approach for modeling mechanisms in Sect. 6.2.1: Casini et al. suggest to model a mechanism by an RBN's network variables, whose values are BNs that model the possible states the represented mechanism can be in. In Sect. 6.3 I have shown that the RBN approach has to face three serious problems: It (i) does not feature a graphical representation of how a mechanism's input and output are causally connected to the mechanism's causal micro structure, which is crucial for mechanistic explanation of certain input-output behaviors of the system of interest as well as for providing answers to questions concerning manipulation and control. It does (ii) not allow for representing manipulations of micro variables, and thus, of bottom-up experiments, by means of intervention variables. The RBN formalism tells us that manipulation of a mechanism's micro variables by means of intervention variables will not affect any one of its macro variables (including the mechanism's output variables), which stands in stark contrast to scientific practice and to what scientists regularly observe. The RBN approach (iii) does also not allow for an error free computation of post intervention distributions when representing interventions as arrow breaking interventions.

In Sect. 6.4 I presented an alternative approach for modeling mechanisms by means of causal nets: the MLCM approach I developed in Gebharter (2014). Contrary to Casini et al. (2011), I suggest to model mechanisms by certain subgraphs. This is much closer to the understanding of a mechanism as an input-output function and does not lead to problems (i)–(iii) Casini et al.'s approach has to face. If the MLCM approach gets everything right and actually is able to provide adequate models for mechanisms, then this seems to have some interesting consequences for the new mechanist movement in the philosophy of science: Providing mechanistic explanation and prediction as well as giving us answers to questions about what can be achieved by mechanistic manipulation and control can be done by specifying the causal relationships between a mechanism's input and output variables in more detail; learning these relationships is the only thing relevant for these purposes. Then, of course, the whole discussion about constitution and constitutive relevance is more or less superfluous and providing mechanistic explanation, prediction, etc. is, basically, nothing over and above providing causal explanation, prediction, etc. We then could say that to mechanistically explain a certain input-output behavior is to explain this behavior by uncovering the system of interest's causal micro-structure that produces this input-output behavior. Throughout Sects. 6.3 and 6.4 I have also discussed Casini's (2016) response to Gebharter (2014).

In Sect. 6.5 I investigated a novel approach to represent constitutive relevance relations in mechanisms. I argued that constitutive relevance relations share one important formal property with direct causal connections: They produce the Markov factorization. Because of that both kinds of relations can be represented by the arrows of a Bayes net. If this argumentation is correct, then this gives rise to a new research area. The question arises if and how already established search algorithms for causal structures can be used to uncover constitutive relevance relations. To give a first sketch how this might be done, I showed how PC could be used to uncover constitutive relevance relations in mechanisms if one assumes CMC and CFC. I also made a suggestion how constitutive relevance could be distinguished from causation in the structures PC outputs. The whole approach, however, leads to a number of still open problems and questions. First of all, it is not clear whether there are special limitations of CMC and CFC in case of systems featuring causal as well as constitutional relationships and, if there are such limitations, whether there is some kind of workaround. If taken seriously, then the approach also seems to lead straightforward into some kind of macro epiphenomenalism. As a consequence, it is not only impossible to intervene on macro variables supervening on micro variables, it is also impossible for such macro variables to be causally connected to any other variable. So if my representation for constitutive relevance relations is correct, then there seems to be no mental causation (see also Gebharter 2015), but also no macro causation in general. In addition, there would be no inter level causal relations. Actually, all the causal work would be done by variables describing the fundamental physical level. These consequences have to be explored in detail.

As a last point I showed how the MLCM approach can be further developed in such a way that it can also be fruitfully applied to mechanisms featuring causal feedback (Sect. 6.6). I used Clarke et al.'s (2014) distinction between static and dynamic problems and followed their example of solving static problems by means of endorsing Pearl's (2000) notion of d-separation instead of CMC and solving dynamic problems by rolling out the causal cycles over time. The paths in the models I use to solve these problems are, contrary to Clarke et al.'s equilibrium networks and dynamic causal Bayes nets, intended to represent the "true" causal structure of the modeled mechanisms. They can, hence, not only be used for generating predictions based on observation, but also for causal explanation as well as for predicting the effects of possible interventions.

Acknowledgements

This work was supported by Deutsche Forschungsgemeinschaft (DFG), research unit Causation, Laws, Dispositions, Explanation (FOR 1063). I would like to thank my Ph.D. supervizor Gerhard Schurz for his great support, for the many possibilities I had during my studies, for so many important and interesting discussions, and for the great time I had over the last years at the Düsseldorf Center for Logic and Philosophy of Science (DCLPS). I would also like to thank Markus Schrenk for his constant support and helpful feedback on many of the ideas and arguments which can be found in this book. My work over the last couple of years also greatly benefited from discussions with Andreas Hüttemann, Clark Glymour, Christopher Hitchcock, Michael Baumgartner, Jon Williamson, Marcel Weber, Frederick Eberhardt, Jim Woodward, Stathis Psillos, Carl Craver, Paul Näger, Bert Leuridan, Erik Weber, Theo Kuipers, Markus Eronen, Lorenzo Casini, the members of the DFG funded research unit Causation, Laws, Dispositions, Explanation (FOR 1063), and my colleagues at the DCLPS. My sincere thanks to all of them. Last but not least, I am indebted to an anonymous referee for incredibly helpful comments on an earlier version of this book and to Otávio Bueno for considering it to be published in the Synthese Library book series.

© Springer International Publishing AG 2017
A. Gebharter, *Causal Nets, Interventionism, and Mechanisms*,
Synthese Library 381, DOI 10.1007/978-3-319-49908-6

References

Balzer, W., Moulines, C. U., & Sneed, J. D. (1987). *An architectonic for science*. Dordrecht: Springer.

Baumgartner, M. (2009). Interventionist causal exclusion and non-reductive physicalism. *International Studies in the Philosophy of Science, 23*(2), 161–178.

Baumgartner, M. (2010). Interventionism and epiphenomenalism. *Canadian Journal of Philosophy, 40*(3), 359–383.

Baumgartner, M., & Casini, L. (in press). An abductive theory of constitution. *Philosophy of Science*.

Baumgartner, M., & Gebharter, A. (2016). Constitutive relevance, mutual manipulability and fat-handedness. *British Journal for the Philosophy of Science, 67*(3), 731–756.

Bechtel, W. (2007). Reducing psychology while maintaining its autonomy via mechanistic explanation. In M. Schouten & H. L. de Jong (Eds.), *The matter of the mind: Philosophical essays on psychology, neuroscience, and reduction* (pp. 172–198). Oxford: Blackwell.

Bechtel, W., & Abrahamsen, A. (2005). Explanation: A mechanist alternative. *Studies in History and Philosophy of Biological and Biomedical Sciences, 36*, 421–441.

Bechtel, W., & Richardson, R. C. (2000). *Discovering complexity: Decomposition and localization as scientific research strategies*. Princeton: Princeton University Press.

Beebee, H., Hitchcock, C., & Menzies, P (Eds.). (2009). *The Oxford handbook of causation*. Oxford: Oxford University Press.

Blalock, H. M. (1961). Correlation and causality: The multivariate case. *Social Forces, 39*(3), 246–251.

Campbell, J. (2007). An interventionist approach to causation in psychology. In A. Gopnik & L. E. Schulz (Eds.), *Causal learning: Psychology, philosophy, and computation* (pp. 58–66). Oxford: Oxford University Press.

Carnap, R. (1928/2003). *The logical structure of the world and pseudoproblems in philosophy*. Chicago: Open Court.

Carnap, R. (1956). The methodological character of theoretical concepts. In H. Feigl & M. Scriven (Eds.), *The foundations of science and the concepts of psychology and psychoanalysis* (pp. 38–76). Minneapolis: University of Minnesota Press.

Cartwright, N. (1979). Causal laws and effective strategies. *Noûs, 13*(4), 419–437.

Cartwright, N. (1989). *Nature's capacities and their measurement*. Oxford: Oxford University Press.

Cartwright, N. (1999a). Causal diversity and the Markov condition. *Synthese, 121*(1/2), 3–27.

Cartwright, N. (1999b). *The dappled world*. Cambridge: Cambridge University Press.

Cartwright, N. (2001). What is wrong with Bayes nets? *The Monist, 84*(2), 242–264.

© Springer International Publishing AG 2017
A. Gebharter, *Causal Nets, Interventionism, and Mechanisms*,
Synthese Library 381, DOI 10.1007/978-3-319-49908-6

Cartwright, N. (2007). *Hunting causes and using them.* Cambridge: Cambridge University Press.

Casini, L. (2016). How to model mechanistic hierarchies. *Philosophy of Science, 83*(5), 946–958.

Casini, L., Illari, P. M., Russo, F., & Williamson, J. (2011). Models for prediction, explanation and control: Recursive Bayesian networks. *Theoria – An International Journal for Theory, History and Foundations of Science, 26*(70), 5–33.

Clarke, B., Leuridan, B., & Williamson, J. (2014). Modelling mechanisms with causal cycles. *Synthese, 191*(8), 1651–1681.

Collingwood, R. G. (2002). In R. Martin (Ed.), *An essay on metaphysics.* Oxford: Clarendon Press.

Craver, C. (2007a). Constitutive explanatory relevance. *Journal of Philosophical Research, 32,* 3–20.

Craver, C. (2007b). *Explaining the brain.* Oxford: Clarendon Press.

Craver, C., & Bechtel, W. (2007). Top-down causation without top-down causes. *Biology and Philosophy, 22*(4), 547–563.

Danks, D., & Plis, S. (2015). Learning causal structure from undersampled time series. In *JMLR: Workshop and Conference Proceedings*, Hong Kong.

Dawid, A. P. (1979). Conditional independence in statistical theory. *Journal of the Royal Statistical Society. Series B (Methodological), 41*(1), 1–31.

Dowe, P. (2007). *Physical causation.* Cambridge: Cambridge University Press.

Eberhardt, F., & Scheines, R. (2007). Interventions and causal inference. *Philosophy of Science, 74*(5), 981–995.

Eells, E. (1987). Probabilistic causality: Reply to John Dupré. *Philosophy of Science, 54*(1), 105–114.

Eells, E., & Sober, E. (1983). Probabilistic causality and the question of transitivity. *Philosophy of Science, 50*(1), 35–57.

Eronen, M. I. (2011). *Reduction in philosophy of mind.* Heusenstamm: De Gruyter.

Eronen, M. I. (2012). Pluralistic physicalism and the causal exclusion argument. *European Journal for Philosophy of Science, 2*(2), 219–232.

Fagan, M. (2013). *Philosophy of stem cell biology.* Basingstoke: Palgrave Macmillan.

Fazekas, P., & Kertesz, G. (2011). Causation at different levels: Tracking the commitments of mechanistic explanations. *Biology and Philosophy, 26*(3), 365–383.

French, S. (2008). The structure of theories. In S. Psillos & M. Curd (Eds.), *The Routledge companion to philosophy of science* (pp. 269–280). London: Routledge.

Friedman, M. (1974). Explanation and scientific understanding. *Journal of Philosophy, 71*(1), 5–19.

Gasking, D. (1955). Causation and recipes. *Mind, 64*(256), 479–487.

Gebharter, A. (2013). Solving the flagpole problem. *Journal for General Philosophy of Science, 44*(1), 63–67.

Gebharter, A. (2014). A formal framework for representing mechanisms? *Philosophy of Science, 81*(1), 138–153.

Gebharter, A. (2015). Causal exclusion and causal Bayes nets. *Philosophy and Phenomenological Research.* doi: 10.1111/phpr.12247.

Gebharter, A. (2016). Another problem with RBN models of mechanisms. *Theoria – An International Journal for Theory, History and Foundations of Science, 31*(2), 177–188.

Gebharter, A., & Kaiser, M. I. (2014). Causal graphs and biological mechanisms. In M. I. Kaiser, O. R. Scholz, D. Plenge, & A. Hüttemann (Eds.), *Explanation in the special sciences* (pp. 55–85). Dordrecht: Springer.

Gebharter, A., & Schurz, G. (2014). How Occam's razor provides a neat definition of direct causation. In J. M. Mooij, D. Janzing, J. Peters, T. Claassen, & A. Hyttinen (Eds.), *Proceedings of the UAI workshop Causal Inference: Learning and Prediction*, Aachen.

Gebharter, A., & Schurz, G. (2016). A modeling approach for mechanisms featuring causal cycles. *Philosophy of Science, 83*(5), 934–945.

Glauer, R. D. (2012). *Emergent mechanisms.* Münster: Mentis.

Glennan, S. (1996). Mechanisms and the nature of causation. *Erkenntnis, 44*(1), 49–71.

Glennan, S. (2002). Rethinking mechanistic explanation. *Philosophy of Science, 69*(3), S342–S353.

Glennan, S. (2009). Mechanisms. In H. Beebee, C. Hitchcock, & P. Menzies (Eds.), *The Oxford handbook of causation* (pp. 315–325). Oxford: Oxford University Press.

Glymour, C. (2004). Critical notice. *British Journal for the Philosophy of Science, 55*(4), 779–790.

Glymour, C., Spirtes, P., & Scheines, R. (1991). Causal inference. *Erkenntnis, 35*(1/3), 151–189.

Good, I. J. (1959). A theory of causality. *British Journal for the Philosophy of Science, 9*(36), 307–310.

Graßhoff, G., & May, M. (2001). Causal regularities. In W. Spohn, M. Ledwig, & M. Esfeld (Eds.), *Current issues in causation* (pp. 85–114). Paderborn: Mentis.

Grünbaum, A. (1962). Temporally-asymmetric principles, parity between explanation and prediction, and mechanism versus teleology. *Philosophy of Science, 29*(2), 146–170.

Harbecke, J. (2015). The regularity theory of mechanistic constitution and a methodology for constitutive inference. *Studies in History and Philosophy of Science Part C: Studies in History and Philosophy of Biological and Biomedical Sciences, 54*, 10–19.

Hausman, D. (1998). *Causal asymmetries*. Cambridge: Cambridge University Press.

Healey, R. (2009). Causation in quantum mechanics. In H. Beebee, C. Hitchcock, & P. Menzies (Eds.), *The Oxford handbook of causation*. Oxford: Oxford University Press.

Hempel, C. G. (1958). The theoretician's dilemma. In C. G. Hempel (Ed.), *Aspects of scientific explanation and other essays in the philosophy of science* (pp. 173–228). New York: Free Press.

Hitchcock, C. (2010). Probabilistic causation. In E. N. Zalta (Ed.), *Stanford encyclopedia of philosophy*. Retrieved from https://plato.stanford.edu/archives/win2010/entries/causation-probabilistic/

Hitchcock, C., & Woodward, J. (2003). Explanatory generalizations, part II: Plumbing explanatory depth. *Noûs, 37*(2), 181–199.

Hoover, K. D. (2001). *Causality in macroeconomics*. Cambridge: Cambridge University Press.

Hume, D. (1738/1975). *A treatise of human nature*. Oxford: Clarendon Press.

Hume, D. (1748/1999). *An enquiry concerning human understanding*. Oxford: Oxford University Press.

Illari, P. M., & Williamson, J. (2012). What is a mechanism? Thinking about mechanisms across the sciences. *European Journal for the Philosophy of Science, 2*(1), 119–135.

Kaplan, D. M. (2012). How to demarcate the boundaries of cognition. *Biology and Philosophy, 27*(4), 545–570.

Kistler, M. (2009). Mechanisms and downward causation. *Philosophical Psychology, 22*(5), 595–609.

Kitcher, P. (1989). Explanatory unification and the causal structure of the world. In P. Kitcher & W. Salmon (Eds.), *Scientific explanation* (pp. 410–505). Minneapolis: University of Minnesota Press.

Korb, K., Hope, L. R., Nicholson, A. E., & Axnick, K. (2004). Varieties of causal intervention. In *Pricai 2004: Trends in Artificial Intelligence, Proceedings* (Vol. 3157, pp. 322–331). Berlin: Springer.

Lauritzen, S. L., Dawid, A. P., Larsen, B. N., & Leimer, H. G. (1990). Independence properties of directed Markov fields. *Networks, 20*(5), 491–505.

Leuridan, B. (2012). Three problems for the mutual manipulability account of constitutive relevance in mechanisms. *British Journal for the Philosophy of Science, 63*(2), 399–427.

Lewis, D. (1970). How to define theoretical terms. *Journal of Philosophy, 67*(13), 427–446.

Lewis, D. (1973). Causation. *Journal of Philosophy, 70*(17), 556–567.

Machamer, P., Darden, L., & Craver, C. (2000). Thinking about mechanisms. *Philosophy of Science, 67*(1), 1–25.

Mackie, J. L. (1965). Causes and conditions. *American Philosophical Quarterly, 2*(4), 245–264.

Mackie, J. L. (1974). *The cement of the universe*. Oxford: Clarendon Press.

McLaughlin, B., & Bennett, K. (2011). Supervenience. In E. N. Zalta (Ed.), *Stanford encyclopedia of philosophy*. Retrieved from https://plato.stanford.edu/archives/win2011/entries/supervenience/

Menzies, P., & Price, H. (1993). Causation as a secondary quality. *British Journal for the Philosophy of Science, 44*(2), 187–203.

Murphy K. P. (2002). *Dynamic Bayesian networks*. UC Berkeley, Computer Science Division.

Murray-Watters, A., & Glymour, C. (2015). What is going on inside the arrows? Discovering the hidden springs in causal models. *Philosophy of Science, 82*(4), 556–586.

Näger, P. M. (2016). The causal problem of entanglement. *Synthese, 193*(4), 1127–1155.

Neapolitan, R. E. (1990). *Probabilistic reasoning in expert systems*. New York: Wiley.

Neapolitan, R. E. (2003). *Learning Bayesian networks*. Upper Saddle River: Prentice-Hall.

Norton, J. D. (2009). Is there an independent principle of causality in physics? *British Journal for the Philosophy of Science, 60*(3), 475–486.

Nyberg, E., & Korb, K. (2006). Informative interventions. Technical Report 2006/204, School of Computer Science, Monash University.

Papineau, D. (1996). Theory-dependent terms. *Philosophy of Science, 63*(1), 1–20.

Pearl, J. (1988). *Probabilistic reasoning in intelligent systems: Networks of plausible inference*. San Mateo: Morgan Kaufmann.

Pearl, J. (1995). Causal diagrams for empirical research. *Biometrika, 82*(4), 669–688.

Pearl, J. (2000). *Causality* (1st ed.). Cambridge: Cambridge University Press.

Pearl, J., & Dechter, R. (1996). Identifying independencies in causal graphs with feedback. In *UAI'96: Proceedings of the Twelfth International Conference on Uncertainty in Artificial Intelligence* (pp. 420–426). San Francisco: Morgan Kaufmann.

Pearl, J., & Paz, A. (1985). Graphoids: A graph-based logic for reasoning about relevance relations. UCLA Computer Science Department Technical Report 850038. Advances in Artificial Intelligence-II.

Pearl, J., Verma, T., & Geiger, D. (1990). Identifying independence in Bayesian networks. *Networks, 20*(5), 507–534.

Price, H. (1991). Agency and probabilistic causality. *British Journal for the Philosophy of Science 42*(2), 157–176.

Psillos, S. (2009). Regularity theories. In H. Beebee, C. Hitchcock, & P. Menzies (Eds.), *The Oxford handbook of causation* (pp. 131–157). Oxford: Oxford University Press.

Raatikainen, P. (2010). Causation, exclusion, and the special sciences. *Erkenntnis, 73*(3), 349–363.

Ramsey J., Gazis, P., Roush, T., Spirtes, P., & Glymour, C. (2002). Automated remote sensing with near infrared reflectance spectra: Carbonate recognition. *Data Mining and Knowledge Discovery, 6*(3), 277–293.

Reichenbach, H. (1935/1971). *The theory of probability*. Berkeley: University of California Press.

Reichenbach, H. (1956/1991). *The direction of time*. Berkeley: University of California Press.

Reutlinger, A. (2012). Getting rid of interventions. *Studies in History and Philosophy of Science Part C: Studies in History and Philosophy of Biological and Biomedical Sciences, 43*(4), 787–795.

Richardson, T. (2009). A factorization criterion for acyclic directed mixed graphs. In J. Bilmes & A. Ng (Eds.), *Proceedings of the 25th Conference on Uncertainty in Artificial Intelligence*, Montreal (pp. 462–470). AUAI Press.

Richardson, T., & Spirtes, P. (2002). Ancestral graph Markov models. *Annals of Statistics, 30*(4), 962–1030.

Russell, B. (1912). On the notion of cause. *Proceedings of the Aristotelian Society, 13*, 1–26.

Salmon, W. (1984). *Scientific explanation and the causal structure of the world*. Princeton: Princeton University Press.

Salmon, W. (1997). *Causality and explanation*. New York: Oxford University Press.

Schurz, G. (2001). Causal asymmetry independent versus dependent variables, and the direction of time. In W. Spohn, M. Ledwig, & M. Esfeld (Eds.), *Current issues in causation* (pp. 47–67). Paderborn: Mentis.

Schurz, G. (2008). Patterns of abduction. *Synthese, 164*(2), 201–234.

Schurz, G. (2013). *Philosophy of science: A unified approach*. New York: Routledge.

Schurz, G. (2015). Causality and unification: How causality unifies statistical regularities. *Theoria – An International Journal for Theory, History and Foundations of Science, 30*(1), 73–95.

Schurz, G. (in press). Interactive causes: Revising the Markov condition. *Philosophy of Science.*

Schurz, G., & Gebharter, A. (2016). Causality as a theoretical concept: Explanatory warrant and empirical content of the theory of causal nets. *Synthese, 193*(4), 1073–1103.

Shapiro, L. A. (2010). Lessons from causal exclusion. *Philosophy and Phenomenological Research, 81*(3), 594–604.

Shapiro, L. A., & Sober, E. (2007). Epiphenomenalism – The Do's and the Don'ts. In G. Wolters & P. Machamer (Eds.), *Studies in causality: Historical and contemporary* (pp. 235–264). Pittsburgh: University of Pittsburgh Press.

Skyrms, B. (1980). *Causal necessity: A pragmatic investigation of the necessity of laws.* New Haven: Yale University Press.

Sneed, J. D. (1979). *The logical structure of mathematical physics.* Dordrecht: Reidel.

Soom, P. (2011). *From psychology to neuroscience.* Frankfurt: Ontos.

Soom, P. (2012). Mechanisms, determination and the metaphysics of neuroscience. *Studies in History and Philosophy of Science Part C: Studies in History and Philosophy of Biological and Biomedical Sciences, 43*(3), 655–664.

Spirtes, P. (1995). Directed cyclic graphical representations of feedback models. In P. Besnard & S. Hanks (Eds.), *Proceedings of the 11th Conference on Uncertainty in Artificial Intelligence* (pp. 491–498). San Francisco: Morgan Kaufman.

Spirtes, P., Glymour, C., & Scheines, R. (1993). *Causation, prediction, and search* (1st ed.). Dordrecht: Springer.

Spirtes, P., Glymour, C., & Scheines, R. (2000). *Causation, prediction, and search* (2nd ed.). Cambridge: MIT Press.

Spirtes, P., Meek, C., & Richardson, T. (1999). An algorithm for causal inference in the presence of latent variables and selection bias. In *Proceedings of the 11th Conference on Uncertainty in Artificial Intelligence* (pp. 499–506). San Francisco: Morgan Kaufman.

Spohn, W. (2001). Bayesian nets are all there is to causal dependence. In M. C. Galavotti, D. Costantini, & P. Suppes (Eds.), *Stochastic dependence and causality* (pp. 157–172). Stanford: CSLI Publications.

Spohn, W. (2006). Causation: An alternative. *British Journal for the Philosophy of Science, 57*(1), 93–119.

Steel, D. (2005). Indeterminism and the causal Markov condition. *British Journal for the Philosophy of Science, 56*(1), 3–26.

Steel, D. (2006). Homogeneity, selection, and the faithfulness condition. *Minds and Machines, 16*(3), 303–317.

Strevens, M. (2007). Review of Woodward making things happen. *Philosophy and Phenomenological Research, 74*(1), 233–249.

Suppes, P. (1970). *A probabilistic theory of causality.* Amsterdam: North-Holland.

Tian, J., & Pearl, J. (2002). A general identification condition for causal effects. In *AAAI-Proceedings*, Edmonton (pp. 567–573). AAAI/IAAI.

Tomasello, M. (2009). *The cultural origins of human cognition.* Cambridge: Harvard University Press.

Verma, T. (1987). *Causal networks: Semantics and expressiveness.* Technical Report, Cognitive Systems Laboratory, University of California.

von Wright, G. (1971). *Explanation and understanding.* Ithaca: Cornell University Press.

Williamson, J. (2005). *Bayesian nets and causality.* Oxford: Oxford University Press.

Williamson, J. (2009). Probabilistic theories of causality. In H. Beebee, C. Hitchcock, & P. Menzies (Eds.), *The Oxford handbook of causation* (pp. 185–212). Oxford: Oxford University Press.

Williamson, J., & Gabbay D. (2005). Recursive causality in Bayesian networks and self-fibring networks. In D. Gillies (Ed.), *Laws and models in the sciences* (pp. 173–221). London: Oxford University Press.

Woodward, J. (2002). What is a mechanism? A counterfactual account. *Philosophy of Science, 69*(3), S366–S377.

Woodward, J. (2003). *Making things happen.* Oxford: Oxford University Press.

Woodward, J. (2008a). Mental causation and neural mechanisms. In J. Hohwy & J. Kallestrup (Eds.), *Being reduced* (pp. 218–262). Oxford: Oxford University Press.

Woodward, J. (2008b). Response to Strevens. *Philosophy and Phenomenological Research, 77*(1), 193–212.

Woodward, J. (2009). Agency and interventionist theories. In H. Beebee, C. Hitchcock, & P. Menzies (Eds.), *The Oxford handbook of causation* (pp. 234–262). Oxford: Oxford University Press.

Woodward, J. (2011a). Causation and manipulability. In E. N. Zalta (Ed.), *Stanford encyclopedia of philosophy*. Retrieved from https://plato.stanford.edu/archives/win2011/entries/causation-mani/

Woodward, J. (2011b). Scientific explanation. In E. N. Zalta (Ed.), *Stanford encyclopedia of philosophy*. Retrieved from https://plato.stanford.edu/archives/win2011/entries/scientific-explanation/

Woodward, J. (2013). Mechanistic explanation: Its scope and limits. *Aristotelian Society Supplementary, 87*(1), 39–65.

Woodward, J. (2015). Interventionism and causal exclusion. *Philosophy and Phenomenological Research, 91*(2), 303–347.

Woodward, J., & Hitchcock, C. (2003). Explanatory generalizations, part I: A counterfactual account. *Noûs, 37*(1), 1–24.

Wright, S. (1921). Correlation and causation. *Journal for Agricultural Research, 20*(7), 557–585.

Zhang, J. (2008). Causal reasoning with ancestral graphs. *Journal of Machine Learning Research, 9*, 1437–1474.

Zhang, J., & Spirtes, P. (2008). Detection of unfaithfulness and robust causal inference. *Minds and Machines 18*(2), 239–271.

Zhang, J., & Spirtes, P. (2011). Intervention, determinism, and the causal minimality condition. *Synthese, 182*(3), 335–347.

Zhang, J., & Spirtes, P. (2016). The three faces of faithfulness. *Synthese, 193*(4), 1011–1027.

Printed in the United States
By Bookmasters